Computer Applications in Agriculture

Also of Interest

Computerized Literature Searching: Research Strategies and Databases, Charles L. Gilreath

Financing the Agricultural Sector: Future Challenges and Policy Alternatives, Stephen C. Gabriel, Dean W. Hughes, Michael D. Boehlje, and Peter J. Barry

Agribusiness and the Small-Scale Farmer: A Dynamic Partnership for Development, Ruth Karen and Simon Williams

Modeling Farm Decisions for Policy Analysis, edited by Kenneth H. Baum and Lyle P. Schertz

Risky Agricultural Markets: Price Forecasting and the Need for Intervention Policies, Pasquale Scandizzo, Peter Hazell, and Jock Anderson

About the Book and Authors

U.S. agriculture appears to be at a major turning point in terms of technological change and innovation as it enters the information age—and at the heart of the information revolution is the microcomputer. This handbook explains in practical terms how computers are being used in agriculture and analyzes some of the issues surrounding present and potential computer applications. The authors define "agriculture" in the broadest possible terms, including the traditional aspects of farming, the industries supporting agriculture, service bureaus related to agriculture, classroom instruction and youth development, and the rural family and community. Considered are specific ways microcomputers are changing agriculture, the exact nature of these changes, and how agriculturists are currently adapting microprocessor technology to make agriculture more efficient and viable. Also included is a discussion of the computer software and hardware used in agriculture today, hardware and software purchasing strategies for both individuals and institutions, and sources of information on computer applications in agriculture.

William O. Rasmussen is director of Western Computer Consortium at the University of Arizona. **C.T.K. Ching** is director of the Cooperative Extension Service and the Agricultural Experiment Station at the University of Hawaii, Manoa. **Lucille A. Linden** is extension 4-H youth specialist and associate professor in the College of Agriculture, Washington State University. **Patricia A. Myer** is extension home economist and assistant agent-in-charge, Washoe County Cooperative Extension Service, University of Nevada. **V. Philip Rasmussen, Jr.,** is director of Agronomy Computer Laboratory, Utah State University. **Roy S. Rauschkolb** is director of Arizona Cooperative Extension Service at the University of Arizona. **Charlotte B. Travieso** is staff leader of Management Systems, Program Development Evaluation, and Management Systems within the Cooperative Extension Service in Washington, D.C.

Published in cooperation with
the Western Computer Consortium,
University of Arizona

Computer Applications in Agriculture

William O. Rasmussen
C.T.K. Ching
Lucille A. Linden
Patricia A. Myer
V. Philip Rasmussen, Jr.
Roy S. Rauschkolb
Charlotte B. Travieso

Westview Press / Boulder and London

Published in 1985 in the United States of America by Westview Press, Inc.; Frederick A. Praeger, Publisher; 5500 Central Avenue, Boulder, Colorado 80301

Library of Congress Cataloging in Publication Data
Main entry under title:
Computer applications in agriculture.
Bibliography: p.
Includes index.
1. Agriculture—Data processing. I. Rasmussen, William O.
S494.5.D3C66 1985 630′.2′0854 85-8855
ISBN 0-8133-0061-4
ISBN 0-8133-0062-2 (pbk.)

Printed and bound in the United States of America

10 9 8 7 6 5 4 3 2 1

Contents

Figures

Preface

Computers and agriculture are not strange bedfellows. U.S. agriculture is the most technologically advanced in the world. Accordingly, it is only natural that agriculture should be linked with today's focal point of high technology—the computer.

This book is intended as a self-contained reference source of computer applications in agriculture and related fields. The aim is to be exhaustive in scope but realistic in practice, since the field of computer applications is an especially dynamic one. The book's objectives are threefold: (1) to provide a brief introduction to computers in general; (2) to describe the way computers are currently used in agriculture; and (3) to provide a comprehensive list of references to enable readers to keep abreast of the rapidly changing field of computer applications in agriculture.

The term *agriculture* is used throughout the book in the broadest possible sense. It refers not only to the traditional aspects of farming but also includes the industries supporting agriculture, the rural family and community, and the resources usually associated with farming and rural areas (e.g., land, water, forests, and so on).

This book is directed to the professional agriculturalist—anyone involved in agriculture as just defined. Such people include the faculty and students of educational institutions dealing with agriculture, Cooperative Extension Service personnel, government agency personnel involved with agriculture, and farmers and other rural residents (such as 4-H club members) having agricultural interests. Two main factors influence this broad audience appeal. First, by their very nature computer applications are general and far-reaching. Second, the authors' backgrounds cover practically all aspects of agriculture.

The book had its origins in a project funded by the W. K. Kellogg Foundation to assess the feasibility of a computer applications center in the western United States. The authors were members of a regional study team charged with conducting that analysis. Although they have widely differing backgrounds and interests, all have some tie to agriculture through the land-grant university system. The authors also have various degrees of computer expertise. While conducting the feasibility study, the authors visited the thirteen land-grant universities in the western United States to document the type and scope of computer applications activities in agriculture. In addition, they consulted with members of regional computer applications centers for agriculture (north central and northeast) to learn about activities there. Thus, although their major focus has been in the western region, the authors' activities have been national in scope.

The Authors

Acknowledgments

We are indebted to many people for their help and encouragement in support of this project.

The W. K. Kellogg Foundation supported the effort that resulted in this book. We gratefully acknowledge its generosity, encouragement, and assistance. The foundation enabled us to meet, plan, develop, and edit this publication.

The authors would also like to thank the Western extension directors who first approved the project concept of determining the need for a Western computer center. Special appreciation is due to our own extension directors who appointed us as members of the Western Computer Center Feasibility Study Team, who allowed us to continue to pursue this additional project, and whose cooperation made our visits to individual states more than fruitful.

Without the additional cooperation of the state contacts in each of the thirteen land grant universities colleges of agriculture, we would never have been able to gather the information presented here. These contacts organized our visits by arranging interviews with faculty, staff, and administration; they scheduled meeting rooms and provided transportation or shepherding as needed; and—perhaps most important—they made our task more pleasant. We sincerely thank Jim Smith of Alaska; Linda Ffolliott of Arizona; Ed Nissen of California; Mike Williamson of Hawaii; Kirk Steinhorst of Idaho; Duane Griffith of Montana; Dick Garrett of Nevada; Stan Farlin and Tom McGuckin of New Mexico; Gene Nelson of Oregon; Larry Bond of Utah; Tony Wright of Washington; Harlan Hughes of Wyoming; and others in each of these states.

Further useful suggestions and perceptive comments were provided by Art Hussey of the Northeast Computer Institute (NECI), John Schmidt of the North Central Computer Institute (NCCI), and Tom Thompson of AGNET and the NCCI.

Finally, numerous people were instrumental in reviewing the manuscript before publication and lending us assurance or criticism as appropriate—for this we sincerely thank them.

The Authors

1 / Agriculture and the Information Revolution

This introductory chapter contains a discussion of the information revolution facing society in the mid-1980s, which was brought about by the development of the microcomputer. Specific reference is made to how agriculturalists are adapting microprocessor technology to make agriculture more efficient and viable.

Development of Agriculture

One of the authors recently read about a seminar conducted in his city at which the seminar leader discussed subsisting on air instead of eating. This life style clearly would have many advantages, not the least of which would be the elimination of one of society's major problems: the need to produce food and fiber for an expanding world population.

Lois Hundgate and Ralph Sherman (1979) correctly pointed out that agricultural production is not necessarily a proxy for social progress, but progress is seldom made when members of society are continually hungry. They noted that "abundant food is not the determinant of progress, but it does accompany progress. Poverty and short food supplies go hand in hand."

A widely accepted view is that for societies to progress, agriculture must be fairly highly developed. Only then are members of the society able to divert their energies from the production of food and fiber to other types of activities. Generally speaking, an abundance of food is usually accompanied by an abundance of other commodities. For example, the industrial revolution was made possible by efficiencies in agriculture; with less labor required on farms, a substantial migration took place from rural areas to urban areas, creating a source of labor that could be used in industrial production.

The basics of agriculture have not changed since humankind first prepared the ground, planted seeds, cared for crops, harvested, and ultimately consumed the produce. Agriculture in its broadest context, however, includes not only the production of foods and fibers but also the processing of agricultural commodities and their distribution to final users. Agriculture also encompasses many services and factors of production including the manufacture and sale of herbicides, pesticides, fertilizers, farm equipment, and so on. Agriculture in this broad context also takes in the rural family and the rural community. Without these types of social structure, agriculture could not exist as we know it in the Western world. Thus topics dealing with food preparation and nutrition, the use of fibers in clothing, and family life and relationships within the community—all fall within the general scope of agriculture. Finally, the term agriculture encompasses a publicly supported sector of government services and regulations, scientific research and education, and the provision of market news for commercial agriculture.

Although the concept of agriculture has not changed, it has had major shifts in emphasis, perhaps the most important of which was the reduction in human resources required to produce food and fiber. For example, in the early 1800s a farmer could produce enough food for 4 additional people. A century later, a farmer produced food for 8 additional people, and by the mid-1900s, a farmer produced enough food for 16 additional people. In the 1980s, a farmer can produce food for 60 additional people, and, on commercial farms with gross sales in excess of $10,000, one farmer can produce enough food for approximately 200 people. The abrupt reduction in human resources required in the production of U.S. agricultural commodities has resulted mainly from technological changes, such as the adoption of irrigation in arid areas, the adoption of hybrid corn in the grain-producing areas, and the adaptation to mechanization in production. Without such measures, U.S. agriculture would be considerably different from what it is today.

In the mid-1980s U.S. agriculture appears to have reached another historic turning point in terms of technological change and innovation. Specifically, U.S. agriculture and that of the rest of the Western world are entering into an information revolution, at the heart of which is the microprocessor or microcomputer. The specific ways in which microcomputers are changing the face of agriculture and the exact nature of these changes are the topics of this book. We submit that microcomputers will have at least as significant an impact on agriculture as have some of the other technological changes mentioned.

An Information Society

Although many scholars and observers have commented upon the information revolution, one of the most succinct and powerful statements has been offered by John Naisbitt in *Megatrends* (1982). Naisbitt observed and forecast a society deluged with information:

> We have for the first time an economy based on a key resource that is not only renewable, but self-generating. Running out of it is not a problem, but drowning in it is. For example:
>
> - Between 6,000 and 7,000 scientific articles are written each day.
> - Scientific and technical information now increases 13% per year, which means it doubles every 5.5 years.
> - This rate will soon jump to perhaps 40% per year because of new, more powerful information systems and an increasing population of scientists. That means the data would double every 20 months.
> - By 1985, the volume of information will be somewhere between four and seven times what it was only a few years earlier.

Naisbitt's forecast of the amount of information that will be available is astounding, but it is only part of the story. Although having large amounts of information might be initially viewed as a positive phenomenon, unorganized information is of limited use and is clearly more a liability than an asset. If society is to manage its wealth of information effectively, it must use the new technology inherent in microprocessors. In the simplest sense, microprocessors will allow us to sift through large volumes of information and select only relevant items. Whereas authors and editors once controlled what the public read, large data bases (collections of related information) accessible by microcomputers now permit the individual to select and control what he or she wishes to read.

Naisbitt pointed out that the development of microprocessor technology will occur in three main stages. In the first stage, microprocessor applications will be utilized in devices built around one or more microprocessor chips. Specific examples include the use of microprocessors in video games, toys, calculators, and watches. The second stage involves replacing other technologies, which traditionally have performed industrial tasks, with microprocessors. Carburation in an automobile is being replaced by fuel injectors controlled by microprocessors; typewriters are being supplanted by word processors that are microprocessor controlled. The list seems endless. The third stage of microprocessor technological development involves applying the technology in entirely new ways. Quite often these new applications will grow out of the technology itself. Specific examples of such directions can only be imagined; perhaps some may be in areas of health and medical care.

Agriculture and the Information Revolution

To get a feel for how agriculture is part of the information society, we will take a brief look at the historical evolution of agriculture. Each U.S. farmer now produces food and fiber for approximately fifteen times as many people as did one farmer in the early 1800s. The total land in farms peaked in the 1950s and 1960s at about 1.2 billion acres. The number of farms peaked earlier, in the 1930s, at about 6.8 million farms. Today approximately 2 million farms operate on approximately 1 billion acres. Fewer farms, using less labor, are producing more food per person than ever before.

Agriculture is one of the few remaining industries in which many producers sell in competitive markets because pricing is often determined by supply and demand. On the other hand, producers put their inputs into markets that are not so close to being perfectly competitive. Agriculture is often characterized as having prices for what it produces set under free market conditions and costs set under imperfectly competitive conditions. This situation has resulted in a cost/price squeeze, in which prices have not risen as rapidly as costs. Thus, to survive, agriculture as business must make every effort to become more efficient. Efficiency, and hence lower costs of production, can result from the adoption of new technology. In the mid-1980s, efficiency must come not only from labor-saving devices but also from devices that will make the management of agriculture resources more orderly.

To remain in business, today's agricultural producer must have the best possible information on growing crops, controlling disease, and applying fertilizer, pesticides, and herbicides. In addition, the producer must keep accurate finanical records for making decisions as well as for income tax purposes. The producer also must have viable marketing strategies that recognize the uncertainties of selling crops well before harvesting. To secure pertinent information, which is available and voluminous, the farmer needs microprocessor technology by which to search large data bases and select items significant to a specific operation. Several examples illustrate this point.

1. Above 98° F, chickens and turkeys have a hard time surviving. It takes about two days to prepare houses to keep birds cool during a prolonged hot spell; otherwise they will die from heat stroke. Having

advanced weather information gives farmers lead time to prepare their facilities to combat adverse weather conditions.

2. Drying conditions are critical to the successful harvest of some crops. Hay needs three days of dry weather after cutting or it rots in the fields. Again, advanced weather information is useful.

3. Insects grow or die depending on weather conditions. With decreasing pesticide use it is more important that chemicals be applied to a crop at the right time. By watching the weather, producers know when to spray, allowing for the fact that some sprays have a usefulness of only a day or two.

4. In the livestock feeding business, profitability depends largely on the critical relationship between livestock prices and feed prices. Practically all livestock feeders must know these relationships and integrate them into day-to-day decision-making. In areas where many feedstuffs are available, feeders need to assemble all price information, relate it to livestock price information, and develop least-cost feed rations. If the number of feedstuffs is large, the problem becomes extremely complex. Microprocessors can be used to analyze these types of data and help feeders make profitable decisions.

Before producers can hope to make profitable management decisions, they must have relevant information on the farming operation. This involves record keeping for all phases of the operation: not only performance records on crops and livestock, but also physical inputs and prices in the production process. Since this information in itself can be voluminous, microprocessor technology is extremely important in managing it.

Fig. 1.1. An Extension specialist explains to an Iowa farmer how computers can assist with farm management operations.

Fig. 1.2. Farm accounting and general ledger made easier with computer assistance.

Fig. 1.3. This 4-H member shows that even young teenagers are able to utilize computers.

2 / Computer Concepts and Terms

Whenever one learns a new discipline, whether it is physics, economics, or a foreign language, new words and phrases must be defined and understood. Learning to be an effective microcomputer user is no exception. The purpose of this chapter is to define and explain essential computer terminology, minimizing the number of terms and concepts introduced and the complexity of the definitions. A description of the major parts of the computer system and their functions forms the bulk of the chapter content, supplemented by the discussion of a few additional terms. The reader is referred to Appendix A for a glossary of computer terms and their definitions.

Components of a Computer System

Although talking about a computer seems natural, talking about a "computer system" is more useful. In this book, the terms *computer* and *computer system* are used interchangeably.

Hardware

Effective computer use does not require a detailed understanding of computer components, but the user does need some elementary knowledge of the major parts of a computer system and the functions they perform (see Chapter 4 for a detailed discussion of computer parts). The entire physical computer system is known as *hardware,* a term that refers to a set of physical/electronic components.

Basic computer components are illustrated schematically in Figure 2.1. The processor (also called microprocessor or central processing unit, CPU) is a tiny electronic marvel. Technically, it is an integrated circuit or chip containing a large number of on-off switches etched in silicon. (For a more detailed description of processors, see the first section of Chapter 3.) Processors perform three main functions: (1) They manage the flow of information through the computer system; (2) they perform arithmetic operations such as adding numbers; and

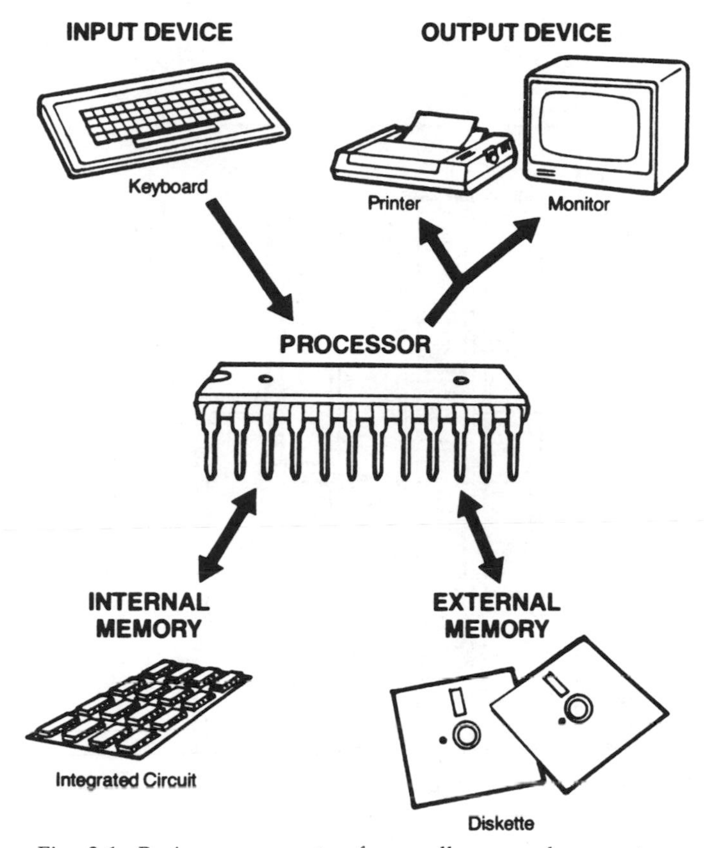

Fig. 2.1. Basic components of a small personal computer.

(3) they execute logical operations of the type "if A is greater than B, multiply A by B." Although these functions seem commonplace, processors are extraordinary in at least two senses: They perform operations rapidly (e.g., they can add two numbers in one-fourth of a millionth of a second); and they are relatively small, at least compared to earlier computer processors. On a typical microcomputer, the processor is about the size of a domino.

Before information can be manipulated by the computer it must be made available to the processor through some form of input device. Most microcomputer systems have a keyboard through which a user can key in or send information to the processor. The microcomputer keyboard is very similar to that of a typewriter. However, because the computer is used to perform much more complicated functions, it has more keys, some of whose jobs are more complex than simply duplicating a number, mark, or letter (see Figure 2.2).

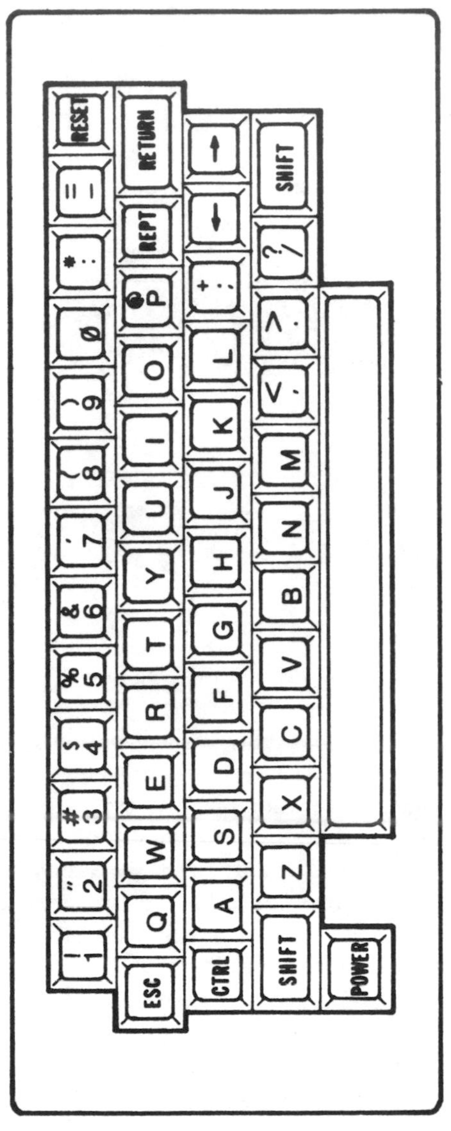

Fig. 2.2. Typical personal computer keyboard.

The computer keyboard has separate keys for the letter *l* and the number *1,* and for the letter *O* and the number *0,* and the user cannot interchange them despite the apparent similarities (e.g., the computer will not add the letter *l* and the number 2). The keyboard also has specialized keys not found on the standard typewriter. For instance, it contains a key labeled ESC, which stands for "escape," and a key labeled CTRL, which stands for "control." When used in conjunction with other keys, the ESC and CTRL keys form additional characters that can be used to communicate with the processor (for a more detailed description of the uses of these keys, see Control Characters section in this chapter). Other keys found on the keyboards of various brands of computers include LINE FEED, left- and right-facing arrows, arrows facing up and down, DELETE, and BREAK. Some keys on the microcomputer have more than one purpose. For instance, when the user keys in text, pressing the RETURN key causes the cursor (or position indicator) to move to the beginning of the next line; this function is similar to that of the carriage return on a typewriter. However, when the user is giving commands to the processor (such as SAVE, PRINT), pressing the RETURN key indicates the end of the command, and the user thus passes control to the computer to act.

When information has been keyed into the computer, the processor can operate on it in an arithmetic sense or store it for some future use. Computer systems have two major types of information storage: internal memory and external memory. The internal memory, usually located on the circuit board of the processor, looks like the other integrated circuits in the computer, with miniature on-off switches etched in silicon. This type of memory is usually volatile—when the computer is turned off, whatever was stored in memory is lost. In contrast, the external memory on a microcomputer system typically is a magnetic disk on which information is recorded in the form of small magnetized spots. The disk is linked to the processor through a disk drive, which writes or reads information on the disk's surface. These devices are comparable to magnetic tape (the disk) and a tape deck (the drive) on a home stereo system. (For a detailed description of disk development, see Chapter 3, Disk Compatibility section.) When the computer system is turned off, the information stored on the magnetic disk is retained and can be used at some future time. The user can either write programs on blank disks or purchase disks that have prepared programs already recorded on them.

Once information is processed by the microcomputer, it must be put into a form that the user can comprehend and use. First, an image can be displayed on a TV-like monitor, on which the user can

Fig. 2.3. The operating system for many small computers is contained within the electronic components on a board or card such as the Z-80 CP/M board for an Apple II® or II+ computer.

Fig. 2.4. A buffer stores information from a computer that is to be printed and then slowly sends it on to a printer, thus freeing the computer to do other activities. This Apple® 64,000-byte serial printer buffer is typical.

see what has been keyed in, what has been retrieved from storage, or what results the computer has reached from performing some operation on the input. On many computers the user can then signal a printer to produce a printed page (printout) of what appears on the monitor. The monitor and printer are examples of output devices.

Software

When one uses a microcomputer, one is actually telling the computer to perform or execute a particular set of instructions. This sequence of instructions is called a computer program, often referred to as software (for a more detailed description of computer software, see Chapter 3). Computer programming is writing directions for the computer, instructing it to perform a specific task. Fortunately, the user does not have to write a computer program every time the computer is used to perform a task; many programs have been written for common tasks and a majority of these programs can be purchased. Generally such commercially available software is more efficient (and probably less costly) than the programs that novice programmers write themselves.

Additional Terms

Statements/Commands. An important distinction should be made between the *statements* in the computer program (the individual instructions) and the *commands* to the processor to do something with the program (for example, store instructions externally or execute the instructions). The program statements entered from the keyboard are not carried out immediately but at a later time. In contrast, direct commands can be given to the processor to do specific tasks, and these are performed immediately. For example, when all program statements are entered into internal memory, the user might issue the command to the processor to SAVE the program in the external memory of the computer. This command provides a permanent record of the program that the user can retrieve at any time. Another command that can be issued by the user is RUN. This directive tells the processor to execute the program in the internal memory of the computer. The computer responds with displays on the monitor or screen telling the user to enter data from the keyboard. Subsequent instructions, when executed, complete the intended task.

Bits and Bytes. The smallest unit of information stored by a computer is a *bit.* A combination of eight bits makes up a *byte*—the amount of memory needed to store one character (number, letter, other symbol). When referring to memory size or disk storage size, the unit *kilobyte* is often used; a kilobyte is composed of 1,024 bytes.

Control Characters. The CTRL (control) and ESC (escape) keys are used to create additional characters to be sent through the keyboard to the processor. When the CTRL key is pressed, no symbol is displayed on the screen. However, when it is held down and another key is pressed simultaneously, a character is sent to the processor although still no character is displayed on the screen. Control characters are especially useful in word processing (see Chapter 3); selected characters appear in the text displayed on the monitor to indicate a specific function (for example, set tabs, print in boldface type, underline) but do not appear in the text when printed. The ESC key is used in a similar manner.

Boot. A common term used with computers is *boot.* This relatively complicated function is performed when one begins to use the computer for a given session. The user must first boot the operating system (see next definition), which in turn allows one to boot an externally stored program and begin utilizing it. It is sufficient to think of the booting process as loading a program into the internal memory of the computer and beginning the directions stored in the program.

Operating System. A computer term used in the preceding paragraph—*operating system*—requires some explanation. When a command is issued to the processor, such as SAVE, RETRIEVE, or RUN, a separate computer program called the operating system is invoked to carry out the command. This system usually resides in external memory and is booted by the user at the beginning of the computer work session. Since many microcomputers use magnetic disks for external storage, operating systems are usually referred to as disk operating systems or DOS.

Disk Formatting. Disk formatting is the placing of certain index marks on the disk. These marks are later used by the operating system and disk hardware when writing information to the disk and when reading from the disk. Disk formatting is roughly like placing lines on a parking lot to assist people in parking their cars; different computers often use different patterns to store information on their disks—they paint their parking lots in a different pattern.

3 / Software

The evaluation and selection of software is the single most important aspect of a computer purchase. However, it is usually the last considered, or even worse it is ignored until after the purchase is made. In this high-tech age, we are frequently motivated in our buying decisions by what is "newest" or "best" in hardware; we justifiably want to buy the latest development in order to prevent premature obsolescence. However, our purchasing decisions often would be much more productive if we were to evaluate the software first, then choose the computer that would best utilize the chosen software. Chapter 5 discusses specific strategies for software and hardware purchases.

A View from Inside the Computer

A computer will not work without proper software: This simple statement cannot be overemphasized. In addition, a computer is only as good as its software: Without proper software, the most powerful, logically complex computer is simply a useless pile of transistors, resistors, capacitors, integrated circuits, wire, and awe-inspiring blinking lights. A digital computer is nothing more than an organized set of on and off switches. The system can only "see" or understand two logical conditions, on or off. These conditions are often represented by a zero (logical off or no) and one (logical on or yes). The computer is really only interpreting whether voltage signals are present or not.

Software controls the way the hardware interprets the voltage state "seen" within the computer and controls the reaction to what is seen. By making logical tests and reactions successively more complex, the computing machine appears to think and perform complex mathematical calculations and data manipulations, responding to us in our own language. One of software's purposes is to insulate us from the mundane, repetitive on-off actions within a computer. By building levels of software from the simplest (0 and 1 codes) to the most complex (our language), we provide a means of translating our commands into the zeros and ones understood by the computer. As we provide more layers of software and progress to higher-level languages, we make our interaction with the computer easier.

An Illustrative Computer Program

To illustrate a simple computer program and to reinforce understanding of the functions performed by computer components (described in Chapter 2), consider the following task:

- Using the keyboard, enter the temperature in degrees centigrade.
- Verify correct entry of the temperature by displaying it on the screen or video display monitor.
- Perform the appropriate arithmetic operations to convert the temperature from degrees centrigrade to degrees Fahrenheit.
- Display the temperatures on the screen with appropriate labels to distinguish the two expressions of temperature.

This example is not presented to show the power of the computer, since the task in question can easily be done with paper and pencil or a pocket calculator. It simply illustrates the functions of a computer and shows how a computer program can be used as a tool to solve problems.

A microcomputer program can be written in a number of computer languages. With the low-level languages, the computer is able to use the instructions directly. This type of computer language is called machine language and consists of a series of zeros and ones (binary number system). Machine languages are not readily intelligible to humans. At the other end of the spectrum are high-level languages that make sense to humans but must be translated or interpreted before they are understood by the processor. Examples of high-level languages include BASIC (beginner's all-purpose symbolic instruction code) and FORTRAN (formula translator).

A computer program written in BASIC to perform the temperature conversion task is shown in Figure 3.1. This computer program is entered from the keyboard and sent to the central processing unit (CPU) where it is directed to the internal memory of the computer. At this point the user has the option of telling the processor either to execute the program or to place the program in the external memory of the computer to permit recall and use at a later time.

Unless otherwise instructed the computer carries out the statements making up the computer program in the sequence in which they

```
10 REM PROGRAM TO CONVERT CENTIGRADE TO FAHRENHEIT
20 REM C=TEMP IN DEGREES CENTIGRADE
30 REM F=TEMP IN DEGREES FAHRENHEIT
40 PRINT "ENTER (TYPE) TEMPERATURE IN DEGREES CENTIGRADE"
50 INPUT C
60 PRINT "DEGREES CENTIGRADE IS ";C
70 F = 32 + 1.8 * C
80 PRINT "DEGREES FAHRENHEIT IS ";F
90 END
```

Fig. 3.1. Listing of the BASIC instructions for the temperature conversion program.

appear; in other words, the instruction in statement 10 is performed before that of statement 20, and so on. The user can give a command that tells the computer that a particular group of instructions, located elsewhere in the program, should be performed out of order, but this case is the exception.

Some features of this computer program illustrate how the computer handles data and interacts with a user. Each statement in the program is numbered; statement 10, for example, is a REMark statement and serves to remind the programmer or user (or anyone who looks at the program) what type of instruction is being issued to the processor. In this case, the REM statement simply tells the user the task to be performed by the computer. Statement 40 is an instruction to the processor to display the phrase enclosed in quotation marks on the video monitor. In this instance the phrase displayed indicates what the user should do to continue the execution of the program.

Statement 50 is an instruction to the processor to display a question mark (?) on the screen. This mark serves as a prompt to the user indicating that the computer is ready to receive information from the keyboard. Once a number is entered it is stored in internal memory and given a name or reference, C. Any subsequent reference to C will involve the number entered in response to this instruction (statement 50). Statement 60 is similar to statement 40. It is an instruction to the processor to display the phrase between the quotation marks and the value of the variable C. Statement 70 instructs the processor to multiply the number symbolized by C by 1.8, add 32 to that product, and call the result F. Statement 80 instructs the computer to display the phrase contained in the quotation marks as well as the temperature in degrees Fahrenheit (F). Finally, statement 90 indicates to the computer that no further instructions will be given and the task is completed.

This sample computer program illustrates both the functions of the components of a computer system and some additional computer terms. The example utilizes a computer program to perform a specific task (granted, a simple one) of multiplying two numbers, adding a number to the product, and displaying the result. The program itself is entered from the keyboard or input device. The information so entered is sent to the processor, where it is subsequently sent to internal memory. Finally, when the RUN command is issued, the video monitor displays the information shown in Figure 3.2 and it can be printed out, if desired.

Computer Languages: From Simple to Complex

The central part of the computer is often called the core because the central memory on older, mainframe (large) computers was referred to as core memory. Software can be divided into the following levels from the computer core to the computer user. Generally, the higher the level of the language, the easier it is for users to write instructions in it.

- The CPU instruction set is the set of instructins built into the CPU itself and is furthest removed from the user.
- Assembly language programs are primitive programs that have to be converted to machine instructions to operate.
- The DOS, or disk operating system, is the control focus of the computer and controls basic storage along with input and output routines. It will operate compiled higher-level and assembled assembly language programs (assembly programs converted to machine instructions).
- Higher-level languages permit interactions with the computer using a language similar to a dialect. The program instructions we write in higher-level languages are called source code.

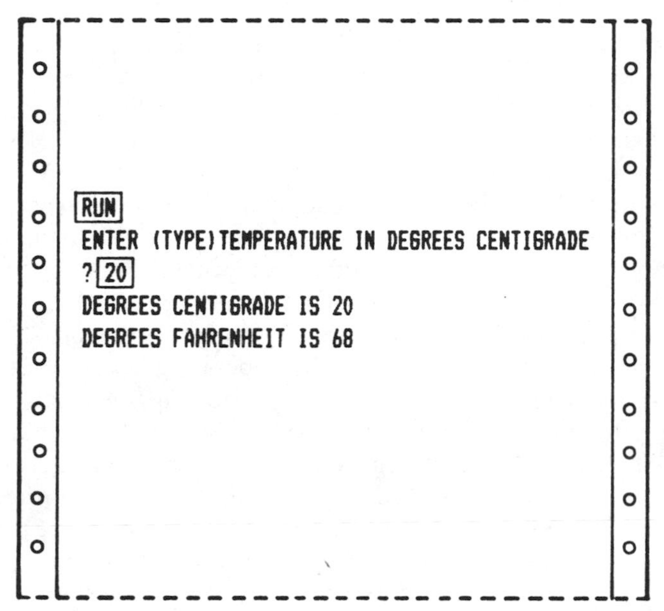

Fig. 3.2. Operation of the temperature conversion program as shown on a computer display screen. Boxed items are those typed by the user.

- Artificial intelligence and program generators are software development tools that can interact with users by asking questions and then generate higher-level language routines to do the job they describe.

Software at the Computer's Core

The first level above the actual logic gates and registers (storage cells) is often never seen by the user and is permanently a part of the CPU. This is the instruction set of the CPU. It is usually in the form of "firmware," permanently engineered into the system when the basic gates and registers are assembled in the factory. Since it is hardwired together, users cannot change it except by choosing a different CPU. For example, potential users of CP/M on the Apple II must buy a separate circuit board with a Z-80 CPU chip before they can use CP/M.

Operating Systems

The next level of software, which stands between the user and the CPU, is the operating system; it may be a simple system for use without external storage devices, a cassette-based operating system (COS), or the currently popular disk-based operating system (DOS).

A DOS usually contains basic routines for converting simple mnemonics of basic logical instructions into specific instruction sets of individual CPU units. It also has basic routines to control the flow of information from the CPU and its closely associated RAM memory (the computer's temporary scratch pad memory) to various peripherals such as modems (connecting devices between computer and communication lines), printers, and other computers. This flow is often called the BIOS (basic input-output system) of the DOS. Another portion of DOS, which controls the flow of information to storage devices, is called the BDOS (basic disk operating system). The DOS serves as the control for most of the basic functions between the CPU and its associated devices. It is the central switchboard and intelligent operator in the flow of information to and from the computer.

In 8-bit computers, eight parallel paths are available for information to move from one component to another. The eight bits that represent a character may thus be sent all at once, one in each of the eight channels, from one computer component to another. An alternate form of transferring the character information would be to use one channel and send one bit after another until all eight bits had been sent; this process would take about eight times as long. A 16-bit machine has sixteen channels connecting each of the components so that two bytes (two characters or sixteen bits) may be sent all at once. This type of machine might appear to be twice as fast at moving data as an 8-bit machine; although this is not exactly the case, it is a good first approximation.

The most popular DOS for microcomputers in 1984 was one of the oldest—the control program/microprocessor (CP/M)—which was developed for the very first commercially available microcomputer, the MITS-Altair. It has survived because it is adaptable: It has been implemented on more than 400 different microcomputers using many versions of the 8080 and Z-80 computer CPU chips. It also has momentum. Because it is available for so many computers, many software manufacturers used its standard in developing their programs. Many high-level languages such as FORTRAN, COBOL, BASIC, and Pascal are available using it as a DOS.

Sixteen-bit IBM PC DOS and its generic equivalent, MS-DOS, are rapidly accumulating programs available for them. *Computers and Software News* forecasts that MS-DOS compatible programs will overtake those compatible with CP/M. One reason is that MS-DOS contains many (but not all) of the same conventions used in CP/M, plus many improvements. Because it is similar (and because the 16-bit 8088 CPU is similar to the older 8-bit 8080 and Z-80 CPUs), older programs that run under CP/M can be converted to MS-DOS. Hence, many popular programs available under CP/M, such as WordStar, dBase II, Abstat, dGraph, and QuickCode, have also been converted to MS-DOS compatible programs. Apple DOS is an extremely forgiving and simple DOS but nonstandard when compared to CP/M and MS-DOS. Other popular DOS systems are the Radio Shack TRS DOS systems of the Tandy-Radio Shack computers. Several TRS DOS versions have many similarities with CP/M.

Some languages used on microcomputers, such as Pascal, have a DOS built into the total package in most versions. (Other versions such as Pascal-MT use CP/M.) UNIX, an older operating system developed for minicomputers by Bell Laboratories, has seen prominence now that 16-bit computers have larger directly addressable RAM memories. Many computer oracles have claimed that UNIX and its associated language, C, will become the operating system of the 1980s. This forecast may or may not come true. UNIX does seem ideally suited to the newest 32-bit microcomputer designs. However, many other recent operating systems such as IBM and MS-DOS 2.+ have many of UNIX's features.

Higher-Level Languages and Applications Programs

Writing programs in assembly language is very time consuming and tedious. It is mainly used when available memory size is critical or when the operation of the program is to be optimal in terms of time and steps used to perform a specific task. Most computer programs are written in a higher-level language. Each language has its own characteristics and features making it well suited for the development, operation, or documentation of specific types of programs. BASIC, an interactive language used for many years, is most likely to be the programming language chosen by the buyer of a microcomputer because it is very user friendly, and a version is available for nearly every existing computer model. It is especially well suited for use on microcomputers and consequently is used extensively. The language FORTRAN is used largely for scientific programming whereas the language COBOL is used heavily for business and accounting operations. Numerous other higher-level languages exist for microcomputers such as APL, Pascal, FORTH, C, and ALGOL.

The major feature of all higher-level languages is the relative ease with which a programmer is able to assemble machine instructions that the computer can read to perform a specific task. Higher-level languages have an associated translator or compiler program that takes the instructions the user has selected in the higher-level language vocabulary and syntax and translates these into machine instructions that the computer can follow. When purchasing a language for a microcomputer the user buys this program along with documentation on the vocabulary and syntax of the language and examples of how to use it. Computer science researchers are currently attempting to produce new languages that will make it easier for the user to convey to the machine what is to be done. Their goal is to allow the user to employ standard English in directing the computer. A program called a translator or compiler will operate on the English instructions to assemble the machine instructions used to drive the computer.

Learning a programming language is by no means required for a person who purchases or uses a computer. A user who needs a program to perform a specific task should look around to see if such a program already has been developed. One needs to look at its availability, ability to operate on a specific computer model, validity, and cost before thinking of writing a program. The monetary equivalence of the time required to write a program can quickly equal the cost of most programs. Many existing microcomputer programs are written in BASIC. A program's BASIC instructions are called the source code. In some cases, the source code is compiled into machine instructions that can be loaded and operated by an operating system. It is relatively easy for a user to change the source code but very difficult to change the compiled code.

The names of several agriculture-related computer software publishers are presented in Appendix B, along with a list of typical software they offer for sale.

Software for Agricultural Professionals

Certain software packages are so versatile and useful that first-time users should consider them carefully. We will discuss the main categories: data base management systems, electronic spreadsheet systems, word processing packages, graphics systems, and terminal communications programs.

Categories of Software

Data Base Management Systems. Because a vast amount of data is available in most fields, the microcomputer user needs to be able to store, retrieve selectively, and display items of possible interest. The storage of such data in a common location and the indexing arrangement that enables users to quickly find them is called a data base. Most computer data bases contain text, numbers, or both. The software program that allows a user to store, recall, or display information from the data base is called a data base management system (DBMS). Situations in which DBMS might be applied include inventory control, customer files, production or service records, personnel or membership directories, and sales or collection records.

A typical DBMS is a substitute for a filing cabinet and manual filing system. Such a system has very few advantages when the information is being entered into the data base: The process is as time consuming and tedious as putting information on paper and into a file drawer. The major advantages of a DBMS are that it is compact (it occupies a small space relative to filing cabinets), and more important, the user can normally retrieve information more quickly and in a more organized form than with a manual filing system. The user can manipulate data in a system by sorting and categorizing and usually can generate reports consisting of organized information presented with labels or headings. Examples include accounting and financial data, name or parts lists, and bibliographic references. Some DBMS packages also have the capability of computing various statistics like totals and averages. Technical computer knowledge is not a prerequisite for using a DBMS.

Electronic Spreadsheets. Perhaps the most novel development and certainly the best-selling packages in the short history of desk top computers are electronic spreadsheets. At some point, every decision-maker uses a piece of paper, be it an accountant's pad, letter tablet, or plain scratch sheet, to perform arithmetic operations on a set of numbers. Some examples are budgets, financial statements and analyses, checkbook reconciliations, investment analyses, income tax calculations, and spending reports. The problem associated with manual worksheets is that once a worksheet is put together it is difficult to correct or make additions or deletions without a large eraser. A small change may require massive recalculations if other numbers are affected (e.g., column totals and row percentages). Recalculations may mean recalling or looking up complex formulas, and they provide an opportunity for simple errors. Finally, these worksheets are normally limited by the physical size of the paper.

Spreadsheet programs for the microcomputer—electronic versions of accounting ledger sheets—circumvent many of these problems. Although several versions are available, they are all basically the same: Most present the user with a computer display divided into columns and rows that will accommodate additions or deletions very easily and also allow rows or columns to be moved around the sheet. One manufacturer provides a spreadsheet that potentially will accommodate 255 rows and 64 columns, or more than forty times the space of a legal-sized accounting pad. Each cell, or space where a row and column meet, can contain an alphanumeric label, a number value, or a formula using not only the usual arithmetic operations but square roots, logarithms, exponents, present value, and trigonometric functions as well. The formula can refer to other cells so that any changes include automatic corrections over the entire worksheet. A key feature is that the user can ask "what if" questions; a single entry can be changed and immediately the effects over the entire worksheet can be seen. As with data base management systems, an individual can easily be an effective user of this powerful class of software without technical computer knowledge. A speadsheet can be as time saving for manipulating large arrays of numbers as a word processing program is for large amounts of text.

Word Processing Packages. One of the more common uses of the microcomputer is word processing. Word processing is very much like typing except that results appear on a video monitor rather than a piece of paper. The advantage of this presentation is that errors can be corrected or modifications made easily on the screeen. For instance, selected portions of long reports can be corrected without retyping the entire manuscript, and line spacing or justification is easily altered. When the user is satisfied with the text, it is simply routed to the printer.

Perhaps the major advantage of word processing systems is speed. The user can type as fast as possible without changing sheets of paper; errors can be quickly corrected by further typing rather than manual erasing and retyping. Multiple copies to different users, sometimes with slight modifications (e.g., form letters), are easily produced. The more elaborate systems have built-in dictionaries that correct spelling and typographical errors. Others can tie in with data bases of mailing lists to put together form letters and mailing labels. Finally, file copies of correspondence or reports can be stored electronically on magnetic disks rather than in file cabinets. Disk storage is often desirable since less physical space is used and since electronic retrieval of information is usually more efficient than manual methods.

Some word processors are relatively easy to use. The basic versions that contain standard features like error correcting, line spacing, and justification require little more than typing skills. Others become more difficult to use as more options (for example, two-column formatting) are included. Once again, technical knowledge of the microcomputer is not required.

Graphics Programs. By graphics packages we refer to computer programs that permit users to display the results of their analyses of numerical data in line graphs, pie charts, or bar graphs. Often a seemingly complicated table of numbers can be shown graphically to emphasize the relationships among the variables. Graphics programs can have spectacular results, especially when shown in color or in three dimensions.

At present, graphics programs fall into two main categories: standalone programs and programs integrated with other application software such as spreadsheets or data base management systems. In both instances they are relatively easy to use. Graphics can be displayed on certain video display monitors (monochrome and/or color), on a printer (usually best on a dot matrix printer), and on specialized plotters.

Terminal Communications Programs. The class of software called terminal programs converts microcomputers into computer terminals (either dumb or intelligent). A dumb terminal has little or no internal memory and merely serves to relay information between a computer and a user. An intelligent terminal has internal memory and a processor. Thus, a microcomputer with terminal communications programs and a modem (a device to attach the computer to a communication line) can be used to communicate with other computers.

Terminal programs for microcomputers have three major uses at present: transmitting electronic mail, giving access to outside data bases, and tapping the power of larger computers. First, the use of electronic mail is rapidly becoming more popular because it combines many of the advantages of telephones and conventional mail without the disadvantages. It is not only very fast relative to the postal or express mail services but also can be less expensive. Further, unlike telephone conversations, both parties can have printed copies of the correspondence. Programs like electronic bulletin boards do not require both parties to be available simultaneously: A user can check the electronic mailbox for incoming messages when convenient.

Second, terminal programs give users access to data bases maintained on large computers. Such bases may contain scientific or financial data or include items of personal interest like entertainment and consumer purchases. Third, terminal programs permit the user to employ the power of large computers for analyses beyond the scope of the microcomputer. For example, entensive mathematical programming or statistical analyses are better suited to large computers because of their speed and memory capacities.

Terminal communications software is potentially the most difficult class of programs to use. If dumb terminal options only are employed, the operation is similar to using other terminals; however, if the microcomputer is used as an intelligent terminal, some fundamental technical know-how may be required.

Integrated Systems. The latest trend in software is to integrate a spreadsheet, DBMS, and word processor (and possibly graphics) into one program so that data can be shared by all areas. Lotus 1-2-3 was the first of these integrated systems. However, a second generation of integrated software is becoming available from the leading industry producers of individual programs. Many have shared-screen windowing capabilities so the user can see portions of each display of the integrated program (such as a spreadsheet) at one time on the screen. Examples of integrated systems are Symphony from the makers of Lotus 1-2-3, Visi-On from the makers of VisiCalc, the Multiplan with Windows from the makers of MBASIC. These packages should make it easier for users to accomplish report-writing functions without large quantities of software and disks.

Program Generators

One of the most popular and powerful data base management programs is dBase II. A user must learn its own rather difficult language in order to operate it and often spend much time programming, testing, and debugging (eliminating errors). The dBase program has a simple, limited text editor that can be used to write operating commands. Typically, a word processing package such as WordStar is used to assemble the commands.

Since many people have trouble writing the instructions to dBase II, program generators have been developed that write dBase programs for a user. A user specifies the characteristics of the data base and the program generates the instructions to be used by dBase II to build it. Several program generators have been developed, such as QuickCode and AutoCode 1. QuickCode, the larger program, has its total files on the distribution disk stored in 229 kilobytes (a kilobyte is roughly equivalent to 1,000 characters); AutoCode is contained in

138 kilobytes. Both programs allow a user to define a data entry form on the computer screen and type in the desired titles, descriptions, or text. The user also can type the name of each data item at the location where it is to be entered. Both programs are well documented, easy to use, and require no dBase programming.

Software Used By Land-Grant Universities

A survey of agricultural professionals gave the following indications of software preference by category. (No endorsements or suggestions are implied.)

Category	*Software Most Often Mentioned*
Word Processing	WordStar (because of availability on many machine types)
Spreadsheets	VisiCalc (an early choice, but many moving to Lotus 1-2-3 or SuperCalc III)
Data base management systems	PFS (an early choice), now dBase II because of availability on many 16- and 8-bit machines
Graphics	Lotus 1-2-3 and many others
Terminal communications	ASCII Express
	Microlink and Microlink II
	Crosstalk
	Z-Term and P-Term

Transportability

We noted that one of the features that led to the prominence of CP/M as a microcomputer operating system is its transportability—the ease with which it can be implemented on many different computer configurations. CP/M can function as long as the computer uses a set or superset of the fundamental 8080 CPU instruction set. However, this requirement limits the machine to certain CPU chips.

An operating-language system that is portable between almost any two computers is the DOS associated with Pascal. This system includes Pascal's cousins ADA, developed and used by the U.S.

Department of Defense, and Modula-2, the follow-on language to Pascal, developed by Pascal's original author, Nickolas Wirth. Rather than having all commands converted directly into machine code, the Pascal language produces an intermediate code; a psuedocode interpreter converts the intermediate code to the specific processor's machine code. Thus almost any processor can run Pascal and its relatives as long as a pseudocode interpreter is installed. These pseudocode interpreters have been developed for most CPU systems.

Disk Compatibility

Disk compatibility is not to be confused with software transportability. Software may be able to work on two different machines, but if disk formats are not compatible, the software is hard to transfer from one machine to the other (see explanation of disk formatting in Chapter 2). The user cannot simply take a disk from one machine, place it in another, and always be able to read it, unless, of course, one is working with two identical computer models.

In early microcomputers, the first floppy disks were disks with 8-inch diameters borrowed from IBM-3740 data entry terminals. Hence, IBM-3740 single-sided, single-density formatting was used on the first disk systems and became a pseudostandard. Most machines could read 3740 formatted disks, even if they primarily used double-sided, double-density techniques to write and read data. If two different computers used 8-inch 3740 format disks, programs or data of similar operating systems (CP/M to CP/M or Pascal to Pascal) could be read from one computer to another by reading each other's disks.

Mini floppy disks (5¼-inch diameter) were developed later and have become the prominent medium for read/write erasable data and program storage. However, most microcomputer manufacturers have chosen to use a unique disk format for mini floppy disk storage. This variety poses a rather difficult problem for universities and agencies that are trying to make software available to other users. Disks from one computer provided to users will seldom operate on another manufacturer's machine. A new microdisk, 3½ inches in diameter, is being used in several computers, including the Macintosh system by Apple Computer, Inc. Unfortunately, at least three different commercial versions are currently available.

With the advent of prominent CP/M computers and the IBM PC DOS (generically MS-DOS), similar formatting was used in many systems. Although disks could not operate directly in machines of other brands, simple software translation routines (such as Crossdata, Media Master, Compate) could make disks readable between many CP/M machines and even across CP/M to PC DOS (MS-DOS). Unfortunately, the Apple DOS disk format was very nonstandard. To date very few ways have been found to read an Apple DOS formatted disk directly on a CP/M or PC DOS/MS-DOS machine.

However, as long as operating systems are the same, information can be transferred between machines of radically different disk formats through RS-232 serial (as opposed to parallel) ports connected directly through wires or over modems and a phone line. The essential item in such a transfer is a data transfer program such as BSTAM, Microlink II, Crosstalk, or Move-it. BSTAM is used primarily for CP/M to CP/M transfer. Others, such as Microlink II, are available in both CP/M and IBM PC DOS/MS-DOS. Thus, because of the similarities in CP/M and PC DOS/MS-DOS operating systems data and programs can be transferred between older 8-bit machines (such as Apple, Osborne, Cromemco, Compupro, and North Star) and newer 16-bit machines (such as IBM, PC/XT, Compaq, Panasonic Senior Partner).

It should be noted that transferring programs between computers is often illegal. Although personally developed data sets or text can be transferred freely, many programs are licensed for use on a single machine or a unique serial-numbered machine. The legal questions involved in this controversial issue are discussed in Chapter 7.

Software Developed at Land-Grant Universities

Surveys of agricultural professionals at land-grant universities have indicated use of a vast array of computer software. However, much of it is special purpose and machine specific. A listing of current software available from these and other U.S. Department of Agriculture (USDA) sources is available from Dr. Robert Strain, University of Florida at Gainesville, for a nominal fee. An updated listing is available as an on-line computer-accessed data base from Virginia Polytechnic Institute and State University.

Examples of programs available by discipline are listed below.

Animal Science
Least-cost ration analysis for dairy, cattle, poultry, hogs
Break-even analysis for cow-calf, stocker calf, feeder calf
Break-even feeder analysis for poultry, hogs, sheep
Dairy economics and purchase analysis

Weight-gain analysis for various animal systems
Herd performance record and analysis packages for dairy, hogs, cattle, poultry
Information packages for health and disease control and analysis

Plant Science
Pesticide and herbicide recommendation packages
Pesticide and herbicide safety packages
Pest and weed growth and development models for optimum spray application and Integrated Pest Management (IPM) systems
Weather data systems for frost warning, pesticide timing
Crop comparisons and analysis for crop choice and management
Crop yield models for management analysis
Greenhouse management and computer control packages
Crop growth models for management

Soil Science and Soil/Water Conservation
Weather and crop-based evapotranspiration irrigation scheduling packages
Evapotranspiration models for irrigation cost/benefit analysis
Soil analysis and fertility test interpretation packages
Soil erosion and crop yield management models
Soil, salt, and water flow and accumulation models for management and control of salinity
Soil chemistry models for nutrient flow and availability analysis
On-line soil testing services
Irrigation management games with graphics
Erosion management games with graphics

Home Economics
Diet analysis
Stored food cost and vitamin/balance analysis
Energy savings analysis for homes, insulation, insulated window dressings
Car cost evaluation
Spending-pattern analysis
Financial-planning analysis
Budget analysis and recommendations
Home-canning cost/benefit analysis
Dieting and weight gain/loss planning
Exercise/caloric-intake analysis

Expanded Food and Nutrition Education Program (EFNEP) records systems

Economics and Accounting
Machine cost analysis
Land purchase analysis
Internal rate of return analysis
Net present value analysis
Machine acquisition rent/buy/custom-work analysis
Set-aside/payment-in-kind (PIK) models and analysis
Enterprise budget worksheets
Accounting systems
Record-keeping systems
Numerous spreadsheet models and analysis for loans, land buying

4-H Clubs and Youth
4-H records systems
Individual club systems (developed by volunteers)
Educational models noted under other headings
Publications management
4-H fair management

Community Resource Development
Community service budget program
Community physical impact model
Various data base systems for state planning use

Natural Resources, Forestry, Hydrology, and Wildlife
Ecological-impact models
Firewood economics
Hydrologic-system models for management
Drought impact models
Wildlife impact models
Ecological educational games for youth and 4-H clubs

Engineering
Machinery management programs
Irrigation planning models
Pumping efficiency models

Pasture and Range Science
Range yield and impact models
Range grazing models
Range pest management models

Extension Administration
Program accountability and evaluation data systems
Budget and salary worksheets and models
Program narrative accomplishment reporting systems

Public Domain Software

A vast amount of software, termed public domain software, is available to anyone for the asking or at most for handling and reproduction costs. Such software comes from many sources; it may have been developed on a government-sponsored project or possibly by a company or individual who then offered it for use by others. The term covers software written in all computer languages and for all operating systems and computers. In some cases a public domain program may perform nearly the same operations as another program sold commercially, and the user can save considerable money by finding out what is available.

Listings of the names and descriptions of public domain software may be found in various software reference books in public or university libraries. A list of CP/M public domain software may be requested from the CP/M Users Group in New York. Another source of information is a local user group for a specific brand of computer. Perhaps the largest and most readily accessible lists are on electronic bulletin boards. Specific bulletin boards contain hundreds of public domain programs available for downloading (transferring a program listing from a distant computer to a user). The telephone numbers of these bulletin boards may be obtained from local user groups or from published books such as the *Computer Phone Book.*

So why buy any commercial software? Much public domain software has only limited documentation on its function, operation, and validation. Although a specific program may do nearly what one wants, it is useless for someone who does not know how to run it. One way around this scarcity of documentation is to ask other users if they know how to operate a program and if it is worth the effort necessary to use it. Another problem with public domain software is that as bugs (or errors) are found in it and corrected by some user, the fixes (or corrections) cannot be distributed to all the software users because they are not included on a central list.

Finally, public domain software may output erroneous values. One may have to check manually the computations done by the program to see if the output is correct. With a purchased program, the output is more likely to be correct because word soon gets around if it is not and the company will not remain long in business. Also, when users of a commercial program report that the program works excellently, the new purchaser is assured that every copy of the program will work the same. With public domain software, a user often does not know if a copy has the same code as a program with the same name being operated by another user.

4 / Hardware

Although computer software has often been termed the soul of the computer, computer hardware is certainly the body, the heart, and the brain. It is easy to become overly preoccupied with hardware: the "gee-whiz" mentality of the blinking lights and buzzing disks can enthrall even the most impartial, hardened professional. New developments, useful or not, often take precedence over the more mundane, proven workhorses of the computer world. The purpose of this chapter is to describe the current technology in such a way as to help the reader make better, more informed decisions regarding which hardware will best fit a particular need.

The western regional computer study team found no simple, easy answers to questions regarding hardware, software, potential growth, compatibility, on-farm uses, software sharing, or the ideal microcomputer. The following questions about the state of the art of microcomputer hardware were commonly raised in each place visited.

- How did technology progress to this point so quickly?
- Are 16-bit or 8-bit microcomputers best for most needs?
- If I have an 8-bit machine, is it outdated and underpowered?
- What microcomputer should I buy, if I am just starting to use a microcomputer?
- How can I get software compatibility? Is CP/M the answer?
- How can I get disk-to-disk compatibility between computers?
- Should I use mainframe computers for most of my work or microcomputers—and how do they mesh together?

The questions multiplied as more states were visited. Directions, trends, and commonalities were postively identified as the interviews continued. Before the state of the art can be outlined, a brief sketch of the development of agricultural computing is in order.

History of Personal Computers

Development of Microcomputer Hardware

As the personal computer industry began to develop (circa 1973–1976), the main agricultural users of the new technology were those persons already involved with electronics in some way. This group included those in the developing market for programmable calculators (statisticians, agricultural economists), those heavily involved with electronics instrumentation (plant physiologists, soil physicists, meteorologists), and those who tinkered with electronics as a hobby. Most microcomputers were available only in kit or semikit form, and an owner's manual read like a textbook for a graduate course in digital design.

In the early 1970s, two computer hobbyists used their common talents and funds to design and build a small, typewriter-sized computer that could be hooked up to a home TV. It had inherent weaknesses: The use of a standard TV as a monitor allowed only a forty-column-wide screen, and the computer terminal had a teletype-style keyboard with only upper-case letters. However, its compact size, color graphics, one-circuit board design, and electronic noise (unwanted electromagnetic radiation) resistance and add-on expansion slots soon led to many sales to friends and several large investors with commercial marketing ideas. The "Silicon Valley computer club" was privy to original design schematics, so as a commercial model was released, an overnight add-on industry was ready to make the little computer far more valuable than it had been in its original form. This open-design philosophy departed drastically from the present-day trade-secret attitudes of large computer vendors.

Eight-bit computers, such as Intertec Superbrain, Apple, Radio Shack, Kaypro, and Commodore, started to appear in Cooperative Extension Service offices. Problems with the 8-bit machines were numerous. Many programs simply would not work in more than one make of machine because of language and operating systems differences.

Meanwhile, an operating system designed to work with many computers—the CP/M—had been developed. Many computers that used CP/M could also use a common computer language, Microsoft BASIC (also called MBASIC). CP/M could also access a simplified set of the common computer languages, such as FORTRAN and COBOL. This combination of a CP/M operating system, termed a disk operating system (DOS), and Microsoft BASIC, CBASIC, FORTRAN, COBOL, and other languages allowed many machines to use identical or very similar programs. There was still no common format for 5¼-inch disks. For 8-inch disks, many microcomputer manufacturers used an old IBM format (3740). Some larger machines with 8-inch disk drives could read each others' programs and data, if CP/M were used as a DOS. Since most small personal computers had no such standard, even if they used CP/M, many problems and frustrations arose.

In 1982, International Business Machines (IBM), the large computer manufacturer, startled the world by announcing that it had developed a small computer that would use an operating system developed by another company. In addition, IBM was reducing security on its proprietary circuit design, making the machine accessible for outside vendor add-ons. The IBM PC would use a 16-bit chip that could directly address large amounts of RAM memory beyond the 64K limit of 8-bit processors. This meant software could be internally complex, yet much simpler to use.

The IBM computer was not perfect. It could not directly run standard 8-bit CP/M software, but CP/M software could be converted to run on it. The IBM PC had a slow computational speed compared to the fastest 8-bit units, and it was expensive and hard to obtain because of extremely high demand. However, its strengths soon made it a leader in sales. In addition, many clones developed that were cheaper, more portable, and more versatile. The clones were different enough to avoid patent infringement but could still operate with many IBM software packages.

Microcomputers in Land-Grant Universities

Because the agricultural software commercial market has many programs for use on IBM or IBM-similar machines, the land-grant universities have turned toward 16-bit machines like the IBM PC. The following list indicates the current trends in microcomputer selection and usage by land-grant university faculties, as determined by a survey by Linda Ffolliott, University of Arizona (1984).

State	*Microcomputers Most Used*
AL	IBM PC; Apple; Radio Shack; Commodore
AK	Apple IIe; IBM PC; Osborne; HP 125
AZ	DEC Rainbow; IBM PC; Kaypro
AR	IBM PC; Osborne; Apple & Apple look-alikes
CA	Compupro 8/16; Kaypro 2; Osborne I
CO	IBM PC; Apple II; Apple III
CT	IBM PC; DEC Rainbow; Hewlett Packard
DE	no response
DC	no response
FL	DEC Professional; DEC Rainbow; IBM PC; Apple
GA	Apple; IBM PC; Vector
GU	IBM PC; Apple; Texas Instruments
HI	IBM PC and compatibles; Apple II+; Osborne
ID	Apple II; IBM PC; Hewlett Packard

IL	IBM PC; Apple II+; DEC PDP 11s; Radio Shack
IN	IBM PC and compatibles; Apple
IA	Apple II; Apple IIe; Apple III
KS	Vector 4; IBM PC; Radio Shack; Kaypro
KY	Radio Shack Models II, III, and 12; IBM PC; Xerox 820
LA	no response
ME	IBM PC and compatibles like Eagle; Apple IIe
MD	Radio Shack model 16; Apple I & II; IBM PC
MA	no response
MI	no response
MN	IBM PC; Apple IIe; Vector 4
MS	IBM PC; Radio Shack; Apple
MO	IBM PC; Apple; Radio Shack
MT	IBM PC; Superbrain; Kaypro
NE	no response
NV	IBM PC; Apple IIe; North Star Horizon
NH	North Star Horizon; Zenith; Apple (usually dedicated)
NJ	Vector Graphic; IBM PC
NM	no response
NY	IBM PC; Apple IIe
NC	Televideo TS-802/803; IBM PC; TRS-80 all models
ND	IBM PC/XT; Apple; Radio Shack
OH	DEC Professional; Radio Shack; Apple; IBM PC
OK	Apple IIe; Radio Shack Model 16; TRS-80 model III
OR	Apple II; IBM PC; DEC Rainbow
PA	Apple II; IBM PC; Radio Shack TRS-80
PR	no response
RI	no response
SC	Radio Shack; IBM PC; Apple
SD	IBM PC; Apple; Kaypro; PET; Hewlett Packard
TN	Radio Shack model 12; IBM PC; Apple IIe
TX	IBM PC; Balcones BNV-205; Kaypro 4
USDA	Wang; IBM PC; Radio Shack III & 16; DEC Rainbow
UT	Apple II; IBM PC; Televideo
VT	Radio Shack; North Star Advantage
VI	no response
VA	IBM PC
WA	IBM PC; Apple II & IIe; Kaypro
WV	Osborne; Apple; Radio Shack
WI	IBM PC; DEC Rainbow; Apple
WY	IBM PC; Apple II+; Apple III

The wave toward 16-bit machines and IBM-similar machines and

IBM-like DOS disk drives has started. We feel that the trend will continue for some time (see, for example, Blundell, 1983; Gens and Christiansen, 1983; Mace, 1984a, Mateosian, 1984; Miller, 1984; Pournelle, 1983; and Zientara, 1984).

Portability

As the computer industry matures, more hardware can be packed into smaller packages; computers the size of wrist watches are forecast. The limiting factor will be human interface with computers. Although more and more power and storage can be packed into smaller and smaller computers, size will continue to be governed by the keyboard and the need for a display that can be easily viewed. The trend toward portability can hold interesting consequences. More work will be done at areas away from the office or office setting, and the results of this new urban movement may be dramatic (Bartimo, 1984).

The emphasis on microcomputers in this book does not show a prejudice against larger multi-user systems; it simply acknowledges that personal computer technology now affects us more directly (Alpert, 1984; Byers, 1984; Gugliotti and Weitz, 1984). Large systems are by no means the dinosaurs of the computer evolution. Some tasks are best performed on a large computer, some on a small microcomputer, and some using both. One of the prime uses of the personal microcomputer will be gaining access to large time-sharing systems for data bases, shopping, information retrieval, and sorting. The utility and power of large computers will be to store, sort, and manipulate data that are too extensive or computationally complex for personal computers. Centralizing large data bases (such as weather data for farmers) in large computers is more cost effective than having each user store redundant information on a personal computer.

Components of a Computer System

Monitors

Most microcomputers with a CRT or video screen have a plug for connecting a monitor. A monitor either is similar to a television set or actually is one. When a television set is used, a radio-frequency (RF) converter typically is required to convert the output from the computer to the input to the television. The RF converter is set to a specific channel number and connected between the computer and the television. If it is set to channel 4, for example, then the television is also tuned to channel 4 and the same display appearing on the microcomputer screen is also presented on the television. In some areas of the country, channel 4 may have a strong television signal so the RF converter is tuned to another channel to avoid interference.

For users who do not like the small screens on many portable computers, a monitor may be added. It also may be used to show a large audience what is occurring on the computer screen. In this case, several monitors may be positioned around a room to allow everyone to have a clear view of one monitor. Other devices allow the image of a computer screen to be projected onto a wall. These are much more expensive than monitors.

Monitors are available with different screen colors; that is, the letters on the screen appear in different colors such as green, orange, black and white (either white letters on a black field or black letters on a white field), and full color (letters may be different colors). The choice of color is usually governed by personal preference; some individuals feel the orange screen is easier to look at for long periods of time without developing eyestrain. Black and white screens are usually felt to be the harshest and hardest to look at for long periods.

Disk Drives

As discussed in Chapter 3, one of the most important developments in the short history of microcomputers has been the mini floppy (5¼-inch-diameter) disk drive. Small, portable, removable, high-density storage for most personal microcomputers had been limited to unreliable and slow cassette tape mechanisms until mini floppy disk drives were perfected in the early 1970s. At that time, many small business and agricultural applications were made possible by this lower-cost storage medium. Although larger floppy disks (8-inch diameter) had been available for some time on units such as IBM 3780 data entry stations, mini floppy disks brought disk data storage to a size compatible with personal computing.

Current technology allows the storage of approximately 100,000 bytes of information on a single-sided, single-density 5¼-inch disk. A single-sided, double-density 5¼-inch disk can contain approximately 250K bytes of data or programs. A double-sided, double-density 5¼-inch disk can hold approximately 500K bytes of data. By comparison, 8-inch-diameter disks can hold approximately double these amounts for similar density and side configurations.

Normally, the higher-density disk drives are preferred, but as byte storage capacity increases, dust sensitivity and other parameters become more critical. Disk failure is more likely in severe environments

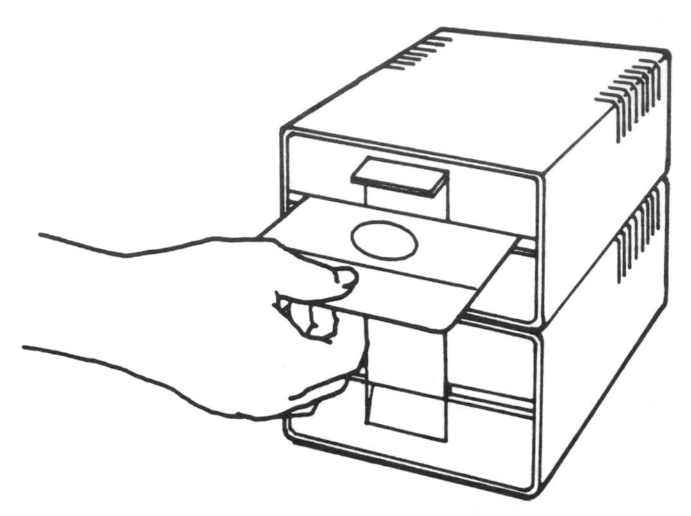

Fig. 4.1. Inserting a diskette into one of two flexible-disk drives.

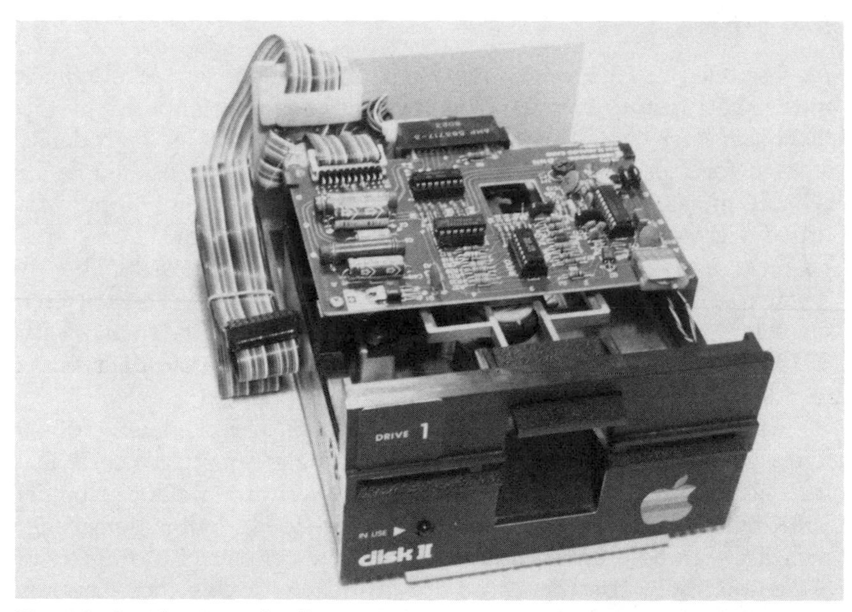

Fig. 4.2. Inside view of a floppy-disk drive used to read or write information on removable diskettes.

or in portable computers. More expensive, higher-quality disks must be purchased for higher-density drive mechanisms. To gain higher storage densities and faster data access speed, the read/write heads must be moved closer to the disks. To achieve this without damaging the disk surface, rigid disks can be made of temperature-stable magnesium alloys and sealed in a pressurized plastic bubble to prevent dust from scratching the disk surface. Such mechanisms are termed hard disk drives. Technological limits on the capacities of these are currently at 5 to 50 megabytes (million bytes) per disk. The advantages to these disks are many, but they are extremely delicate and portable applications are somewhat limited. The user must have a way to make backup copies of the disk in case of mechanism failure such as head crashes. Because hundreds of floppy disks would be needed for this process, other techniques have been developed. Video tape players or cartridge tape drives can be used to make backup copies of hard disks in 10 to 40 minutes.

Printers

Printer technology is currently moving even faster than computer technology. The price of the previously more expensive printers has decreased until it approaches that of the traditionally cheaper types, and this trend will probably continue for the near future. Most users buy a reliable dot matrix printer with graphics capability for general use. Its price ranges from $160 to $2,200, depending on speed of output. Dot matrix printers form characters out of dots of varying sizes. The finer the dot pitch of a printer, the better is its quality, usually at a sacrifice of speed. Speed ranges from 80 to 200 characters per second (cps).

Daisy wheel printers, usually slower, use an impact wheel to form characters much like those produced by expensive office typewriters. They can use carbon ribbons to yield outstanding print quality at speeds much higher than anyone can type. In the past, these printers cost more than $2,000, but recent entries into the market, although somewhat slower, yield similar print quality at prices as low as $395. Speed of the lower-priced units is usually 15 to 25 cps compared with 20 to 100 cps for the higher-priced units.

A third type of printer, which has been out of reach for most personal computer buyers, uses a laser beam to burn an image on an electrostatic drum. The printing process is then similar to a xerographic copier. These printers can produce print quality as good as that by the daisy wheel units at speeds of one page per second and more. Their price has traditionally been $50,000 to $200,000,

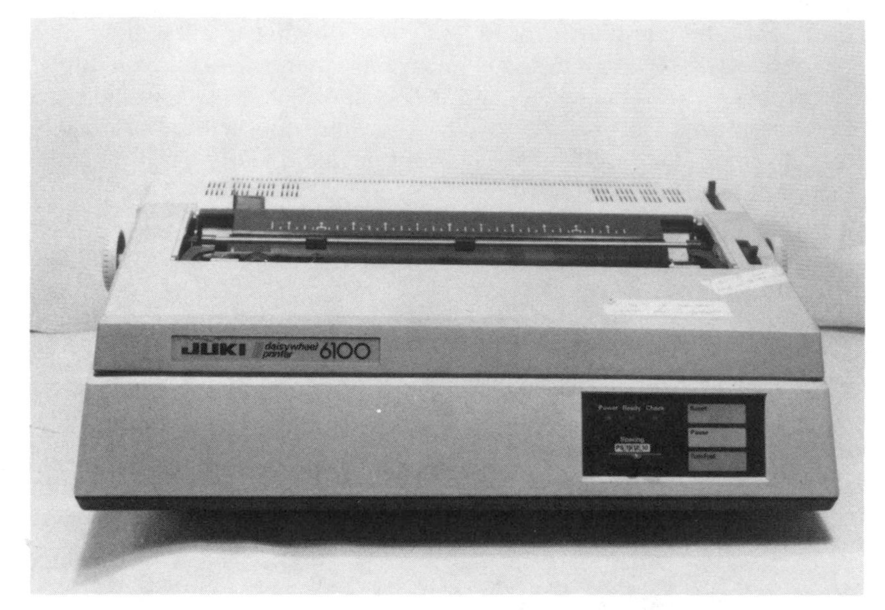

Fig. 4.3. One example of a low-cost, letter-quality, daisy-wheel printer is a JUKI® 6100.

but a recently marketed printer by Hewlett Packard, as well as one by Apple, sliced those prices roughly by a factor of 10. This development may put fast, versatile, quality laser printing in the reach of many agricultural professionals.

Modems

The purchase of a personal computer immediately gives the user the basic unit for communication with other computers. The only items that must be added are a communications package and a modem. A modem (modulator-demodulator) enables a computer user to send data coded in tones over the telephone line. It takes binary codes and translates them to tones (modulates them) and then listens for the tones from another modem and transfers (demodulates) them into binary codes that the computer can understand.

The speed of modem operation has traditionally been limited by computer or terminal speed and the electric noise on normal phone lines. Such noise comes from proximity to other lines and from switching at telephone routing centers. Early modems operated at 110 baud (bits per second) or approximately 10 cps, which was as fast as early terminals (teletype machines) could operate. More recent

Fig. 4.4. An intelligent modem (Hayes Smartmodem®) used for connecting a personal computer and a telephone line. It is controlled by a small internal computer.

modems have used a speed of 300 baud (about 30 cps) because almost any newer video terminal or computer can operate at this speed and even the most noisy phone lines can be used. Recent advances in noise-cancelling electronics have allowed 1,200 baud modems to proliferate and made possible even some 2,400 baud units. Faster modems permit lower total phone charges, of course, but not all telephone systems, especially rural ones, can tolerate 1,200 baud rates. A buyer should check with local computer users and dealers before purchasing the new, lower-priced, high-speed equipment.

For exceptional purposes over special, shielded wires, extremely high-speed modems that reach speeds of more than 100,000 baud, and even higher speeds, are possible. These modems are often used between buildings or offices and as part of local area networks.

An Entry-Level Agricultural Computer System

Many problems are associated with making recommendations for a computer system in such a changeable marketplace. However, a few suggestions frequently heard by the regional study team are valuable for anyone considering a new system for agricultural use. An entry-level system for an agriculturalist or agricultural professional is governed by resources available, intended uses, and need for future expansion. It would probably include the following components.

- A basic 8- or 16-bit CPU computer with appropriate languages and support software for operation. If an 8-bit CPU, it should contain at least 64,000 bytes of RAM memory. If a 16-bit CPU, it should contain at least 128,000 bytes of RAM memory.
- Two disk drives, for proper and necessary backup procedures. These may be two floppy disk drives or one floppy disk drive and one hard disk drive.
- A good quality video monitor. A television set is adequate for playing occasional games, but for repeated professional use, the sharp, clear images of a computer monitor will lessen eyestrain.
- A good quality printer. Many newer dot matrix printers offer excellent printouts. For quality a daisy wheel printer may be desired.
- Adequate power line regulation and protective devices. These may be an inexpensive surge preventer or a battery-assisted backup power supply, depending on need and the value of the data.

More information may be obtained from Osborne, 1979; Segal and Berst, 1982; Meilach, 1982; and Chapter 5 in this text.

Interfaces

Computers normally communicate internally or with each other in one of two ways. The most often used mode is a bit-after-bit serial data communication as described in the section on modems. Messages may be sent on common phone lines as pulses of voltage or in modulated tones. At least two wires are required, the signal wire and a ground. The second type of communication is called parallel data communication and takes place one byte at a time over eight wires, with each wire carrying one bit. This technique is usually faster but requires special shielding for long distances. It is usually used only at distances under 100 feet, such as to printers from a computer or to disk drives from the computer.

There are supposed standards for each of these wiring configurations. However, the only real standard is that they are not standard. One of the most prolific and knowledgeable computer designers has described this problem as follows.

> Among the most exasperating experiences in any computer user's career is connecting two serial devices. I don't mean a terminal and a modem (making that connection is a piece of cake) but any other connection can be real trouble. For instance, every time I buy a new piece of equipment, things seem to go this way: I spend five minutes reading the sales brochure, five minutes reading the financial transaction, and five hours trying to figure out how to make the new equipment communicate with my computer ("Ciarcia's Circuit Cellar," *BYTE*, April 1983).

No easy solutions have been found to solving the problems of interfacing a computer and a peripheral; however, some excellent "cookbook" guides have been written (Campbell, 1984; Marx, 1983). A most productive suggestion for first-time buyers is to buy all computer items from one supplier and have them interfaced by the vendor.

Maintenance and Service

One of the often overlooked areas of a hardware purchase is the cost associated with parts, maintenance, and routine service. Even worse, many users continually neglect service procedures and then curse the machine when a component like a disk drive fails. Seven simple steps could offset 99 percent of all user-caused computer hardware failures. The new user's attention to these areas is critical.

1. See that all equipment has access to the proper voltage outlets. Be sure that outlets are wired and grounded properly; a reversed polarity plug or a poor ground will invite hardware failure in a matter of days. *Never* operate a computer or peripheral that requires a ground by bypassing the ground plug.
2. Static electricity is *extremely* dangerous to the small integrated circuits of a computer. If you feel static sparks crackle in your carpet or clothes, you may do irreparable damage to your computer. Obtain an antistatic spray for the carpet or buy a permanent antistatic mat from your computer dealer.
3. Guard your computer against powerline surges. Inexpensive surge arrestors can be purchased to absorb the transient voltage spikes that accompany most power systems where large loads are turned on and off (e.g., air conditioners, furnace blowers, or welding tools).
4. Keep the computer area as clear as possible. Chalk dust is a slow but sure death to floppy disks. Computers have worked effectively

for years in dusty grain elevators, but users pay a price in worn disks and disk drives.

5. Keep your computer cool. Do not stack paper, books, or other items on it or cover the convection cooling holes. If you have added many peripherals to a computer without a fan, an inexpensive cooling fan is available.

6. Maintain your computer. Disk drive cleaning kits should be used routinely (once a month) to clean the disk read/write heads. Disk drives gradually wear and change speeds. Kits are also available to check disk speed and correct it once the warranty service is expired.

7. If a.c. power is erratic or brownouts are common, consider an uninterruptable power supply (UPS). A unit that supplies 200 watts for 20 minutes of emergency power is available for under $300. This investment can be crucial if your computer has a hard disk. A power outage when the hard disk is writing could erase data representing years of effort.

Guides are available to assist the first time user in proper computer operation and maintenance (e.g., Zaks, 1981; Popenoe, 1984; Ciarcia, 1983; Meilach, 1982; and Stifter, 1983).

Protection Against Loss

Early in the personal computer era, just five years ago, a user had a hard time finding any insurance company, save Lloyds of London, that would insure personal, school, or institutional microcomputers against theft, fire, or loss. Thieves could too easily walk away with transportable units. Insurance is now available from several sources such as Safeware, Inc. Most policies require that computers be attached to tables with protective devices; such attachment devices to deter theft or tampering by students and users are increasingly available.

5 / Microcomputer Purchasing Strategies and Training Approaches

As microcomputer technology advances at a staggering rate, computer users, managers, administrators, and agricultural specialists feel pressure to buy a microcomputer and to know how to use it. They see the rapid proliferation of microcomputers on the farm, hear the advertisements on radio and TV, and often are expected to have first-hand knowledge about them.

In this chapter, we are concerned with how best to plan and execute a computer purchase and how to provide instruction or training on the use of computer technology. We are interested in two perspectives—that of the individual and that of the group of individuals or the organization. (Organizations in this context include business firms and public agencies.)

Purchasing Strategy for an Individual

The *wrong* approach to acquiring a microcomputer is to (1) buy the hardware; (2) identify the applications; and (3) then find appropriate software. Although this chronology may seem laughable to the consumer who thinks logically, microcomputers are nevertheless often purchased by following this traditional approach. The usual result is a frustrated user who, if serious about microcomputer usage, ends up buying additional peripherals or even a whole new microcomputer system at higher cost. The frustrated user also sometimes abandons the whole idea of using the technology and is left with a table full of inactive hardware, a very expensive office decoration or family home night light.

Identify Specific Tasks

The first step for a potential microcomputer buyer is to think about specific needs. What jobs is the microcomputer system being purchased to perform? How can you use it to make your home/office/farm operation run better? List your primary needs and activities and decide which of these could benefit right away from your purchasing a microcomputer. Then think about your secondary needs and try to uncover future uses that you might find for your microcomputer.

Your next step is to look around for software that meets these needs. Visit your local computer stores (they have popped up everywhere) and see some systems in operation. Talk to farmers, ranchers, extension and university specialists, and business people who are using microcomputers. Read books and magazines on your subject area, visit a trade show, attend user group meetings, take a course, seminar, or workshop, do some research and become an informed consumer.

The right purchasing strategy is just a rearrangement of the wrong one: (1) identify the applications; (2) find the appropriate software; and (3) buy the hardware.

Computer Purchasing Alternatives

Lease Equipment. One alternative to buying a microcomputer for your individual use is to lease one for a short time to see what it can do for you. Then you can convert to a purchase once your needs make it worthwhile for you to own a computer. Even though the one you choose to buy may be outmoded in a few years, it will still do what you intended.

Hire a Consultant. If you are not ready to buy a microcomputer or do not want to use one yourself, an alternative solution is to hire a computer specialist who has access to a microcomputer to run the programs you need. Although this alternative can be costly, it can give quick results and can relieve you of having to take the time and effort to buy and familiarize yourself with your own microcomputer.

Alternative Equipment. Not all tasks are best done on a computer. A programmable calculator is much less costly than a microcomputer, is portable, and may be able to handle your needs. Again, this option can be explored with your local dealer or with neighbors, coworkers, or cohorts who use calculators in their businesses or homes.

In some parts of the United States, if you are interested in only one-way communication, for example to access data bases or to get weather information, you may want to explore the purchase of a videotext system. This device connects to your television screen and your telephone. The videotext service allows you access to thousands of electronic pages of updated news, weather, market, and management information at any time of day or night. Some videotext services can give the farmer access to weather forecasts for a 50-mile radius around the farm in the form of full-color maps.

Software Selection

The individual should ask three key questions when selecting good software: (1) Does it meet your needs? (2) Does it do what it says it will do? (3) Does it have good user documentation (easy to read, well-indexed)? Some factors to consider when evaluating and choosing good software for your purposes are the following.

Ease of Use. Is the software easy to use? Is it menu-driven? Does it guide you through the steps of the program?

Help. Where can you go for help if you run into problems? Is there a toll-free phone number that you can call for assistance? Does the package itself have a HELP function through which you can receive on-line assistance? Does the company provide updated versions of the package on a periodic basis? Are they free? Try out the package before you buy it. Some software programs may come with errors in them.

Compatibility. Will the software operate under your computer's operating system? Each microcomputer has an operating system; the most familiar ones are MS-DOS, CP/M, and UNIX. Because software packages are written and designed to work on specific operating systems, be sure that any program that you acquire will run on an operating system that you can use with your hardware.

Memory. Does the computer you are considering have enough memory or storage capability to run your program?

Recommendation. Does the program come from a reputable source? Has it been used and recommended by someone you trust? The best agricultural programs are written by people who combine strong agricultural expertise with the ability to write good user-friendly computer programs. County extension agents may know of programs that have been tailor-made for specific operations, and the land-grant universities may have programs available at nominal cost. Although the quality, quantity, and user-friendliness of agricultural software have improved greatly, programs still need to be evaluated before being selected.

Hardware Selection

When you have identified your needs and the necessary software, your hardware selection should be considerably narrowed. Try to buy everything from the same local dealer, including microcomputer, printer, software, and additional hardware. Check into maintenance agreements and arrangements for service; some dealers may be anxious to sell you a system but not as anxious to fix it if you have problems. You may not want to purchase everything at once but to add peripheral devices or additional memory boards as your needs expand and your finances permit. Make sure that the dealer will be able to accommodate you in the future; a reliable dealer who handles several brands of hardware and software can help you make informed decisions. Some factors to consider when evaluating and comparing hardware are the following.

Memory. How much memory do you need? Most computers come in sizes ranging from 2 kilobytes (approximately 2,000 characters) to 256 kilobytes.

Input/Output Devices. What kind of output do you require? Will you need to keep a printout of the information you generate? Most programs require a printer. Do you need a screen? How big should it be? Some screens can be hard on the eyes.

External Storage. What kind of storage does the microcomputer use: floppy disk, hard disk, or tape? If the system uses floppy disks, does the disk drive come separately? Do you need a disk drive that uses single- or double-density disks?

Training. What kind of training is available on the new equipment?

Service, Maintenance, Support

Once you have bought the computer, will you have adequate backup and maintenance support from the vendor? What happens when the computer is not working or down? Will you have to carry your equipment to the vendor site for servicing? How long will you be without your system? How far away is your dealer? Where will the equipment actually be serviced? It is important to buy something

Fig. 5.1. Additional memory can easily be added to a computer, as on this card containing 256,000 bytes of random access memory (RAM).

Fig. 5.2. Modular expansion cards can be inserted into an IBM-PC for increasing the number of additional cards that can be used. This card contains a clock/calendar and an analog I/O card.

that can be fixed fairly quickly and cheaply. You may want to check the cost and availability of maintenance contracts on your system.

Value. Some vendors offer a "bundled" package in which equipment and software programs are offered together at one price. You need to compare prices carefully, considering the components and software available. The value is not only the price but includes the reputation of the dealer, the reliability of the equipment, and the support, maintenance, and service you will be getting.

References and User Groups

Many microcomputer brands have their own user groups that meet regularly. These groups can provide a wealth of information, both from people who have "been there" and those who are just beginning. In addition, members of user groups can benefit by pooling their resources to buy software or hardware at volume discounts.

The extension office in your county or parish may have a state publication on computers or a checklist for buying. The home economics group in the USDA Extension Service office has put together a one-page fact sheet, "How to Shop for a Home Computer." The land-grant university in your state may be able to direct you to a compiled inventory of software (e.g., such lists have been compiled and updated in Strain and Fieser, 1982).

Purchasing Strategies for Groups and Campuses

Buying microcomputers for a group of users involves a somewhat different strategy from buying a single computer for an individual user. An office, campus, or agricultural group needs to consider the choice between stand-alone microcomputers and multiuser systems. With several stand-alone microcomputers, individual users have control over their own data files, hardware, memory, and software. With a multiuser system, individual users have access, through terminals and telephone lines, to a larger central processing unit. Users can protect the privacy of their data files with passwords and access codes or they can share data files with other users. This sharing can be valuable for keeping calendars, committee rosters, or administrative information. Buying a multiuser system with individual computer terminals can be less expensive than purchasing one stand-alone microcomputer for each user. However, if the multiuser system is fully occupied, access time can be slow, and the core memory can fill up quickly.

Compatibility. If the choice is stand-alone microcomputers, then compatibility is a key issue. Users may want to be able to trade disks or to print files from one system on another. The hardware acquired should permit the users to communicate with larger facilities, such as a university mainframe computer. If computing resources are to be added incrementally, the compatibility factor must be weighed heavily.

In-House Service and Support. With a large group of users, service and assistance are especially important. In some organizations, a computer applications staff or management systems branch is established, consisting usually of a group of computer specialists, technicians, programmers, and other technology experts. The staff operates as a development or service bureau to support the agency's hardware and software.

The Word Processing Dilemma. In many group settings, the microcomputer with a good word processing package can be quickly monopolized for composing, editing, and printing documents. This use of a microcomputer can prove to be inefficient, and care should be taken to provide word and data processing capability for all the users in a group who need it. Depending on usage, some stand-alone word processors can be more expensive than the stand-alone microcomputer with a good word processing package. Some agencies have solved this problem by providing a shared-logic word processing network (a network in which a central computer is tied to several terminals) for secretaries and stand-alone microcomputers with communication capabilities for analysts and researchers.

Resource Availability. Nothing will turn away a novice user faster than a down (not working) computer or a slow shared-access system. Having one piece of equipment for many users can also be a deterrent because users must compete with each other for computer time. Although most organizations are far from providing a microcomputer for every person who needs one, users are more likely to learn to utilize a microcomputer if they do not have to walk up the hall or across the campus.

Effective Training

Effective training is the least-cost and minimum-effort way to provide instruction that will enable the user to carry out a desired task on the computer. The effectiveness of training depends on a number of factors: the motivation of the trainee, the accessibility of computer equipment for the trainee to learn computer use and to practice newly learned skills, the availability of skilled instructors, the sociological/ demographic characteristics of the trainee, and the environment in which the training is presented.

In this section, we provide an overview of computer training activities by distinguishing between different types of training as well as different methods of training. We restrict our discussion to the training most applicable to those interested in computer applications in agriculture and related resources. We make a sharp distinction between a user and an operator because they differ in the degree of their technical training. We view users as people who simply wish to use the computer to perform tasks; they have no interest in learning the technical aspects of computer hardware or software but see computers as problem-solving tools. Operators usually are knowledgeable about computer programming languages and computer systems operating procedures and also have a relatively strong grasp of the technical aspects of the hardware. Because of the objectives and audience for this book, we are clearly more concerned with users than with operators.

Types of Training

Hardware. The objective of hardware training is to prepare individuals to install properly the various physical components of the computer system, to operate the computer system, and to diagnose malfunctions. Because we are emphasizing computer users and not operators, the objectives of hardware training are necessarily very superficial. By install we mean to connect the various components (disk drives, monitors, keyboard, and printer) to form the computer system. By operate we mean to set switches into proper positions and connect components to the correct power sources. By diagnosis we mean to be able to isolate (if possible) and describe the cause of a malfunction. Correct diagnosis can facilitate the repair process, since the user can take the malfunctioning component to qualified service personnel for repair or replacement. For many users, hardware training is covered in the manuals accompanying the computer equipment. Some manuals, for example, have a section on trouble-shooting if the system fails. It is, however, helpful to review these hardware basics in any training session.

Software. Training in software usage is determined by the needs of the user and may or may not include learning computer languages. Because of the number of computer programs available, existing software addresses most general areas and topics. If a user can apply an existing program, learning how to program the computer is not a necessary part of training. Even if the software appears to be expensive, the cost for developing a package can easily go far beyond that of most commercial programs. However, when an extremely specific or unique program is required, the user will need to learn to program in one or more languages, ranging from high-level languages like BASIC to assembly language. For whatever level of software used, documentation needs to be studied by the user whether for operating finished programs or for programming in a specific language.

Methods of Training

To put the notion of computer training in perspective, we distinguish three major types of computer training methods: self-teaching, device-assisted, and direct contact. These categories are neither clearly defined nor mutually exclusive. For example, device-assisted instruction can be used in conjunction with direct contact methods; an effective introduction to computer systems could be showing professionally prepared video tapes (device teaching) and then using the direct contact approach to address specific problems. Each method will be described here as it relates to training users rather than operators.

Self-Teaching. In this training method, the person interested in using a computer simply acquires access to the appropriate hardware, reads the appropriate manuals or other information sources, and performs the intended tasks. The process is clearly one of experimentation and discovery. The user usually accumulates a list of questions to ask a more experienced person when one is available.

This type of training works well for the highly motivated individual who believes in and practices self-teaching and discovery learning. Prerequisites include an inquisitive mind and the ability to search out and read the relevant information. This approach also works best if good, simplified software manuals, geared to self-teaching, are available. This type of training suits those who feel that they do not have to attend a class to learn new material and who take satisfaction in self-teaching.

Device Assisted. Among the many physical devices that individuals can use to learn new materials are audio cassettes related to reading material, audiovisual media involving slide/tape presentations (presentations in which 35-millimeter slides are keyed to audio tape narration) as well as video cassette recordings, and computer-assisted training. These instructional devices are comfortable for those who feel that they must have some external assistance in learning new material. Such individuals also may like using tutorials (intensive teacher-directed instruction) that are regulated by some mechanical or electronic device (e.g., video tape or computer-assisted media).

Of particular interest are the computer-assisted teaching devices, such as PLATO, a system designed to teach computer training as well as other course material like natural and social sciences. Although PLATO has been available for many years on a large mainframe computer in Illinois, selected modules are now available on diskettes for several popular microcomputers. Many software vendors have produced computer-assisted instruction languages to teach materials similar to those presented by PLATO. Other vendors market programs. One popular approach is that prepared by American Training International (ATI), which simulates the actual software being instructed so that the trainee can learn by interacting with the screen.

Direct Contact. Three variations of the traditional student-teacher model are relevant to our discussion: one-to-one contact, one-to-many contact with end users, and one-to-many contact with trainers. One-to-one contact training is conducted by one instructor who teaches one trainee. This variation is, of course, the most intensive of the direct contact methods and the most expensive. It is used quite often in extension teaching, when a specialist or county agent works with an individual producer or rural resident. This type of training is rarely cost effective. A useful variation of one-to-one contact incorporates the self-learning method. One-to-one contact is minimal since the self-learner only works with an instructor on areas where the trainee is having a specific problem. This approach often is a very effective way of transmitting computer knowledge because the subject matter is clearly focused and the learners usually are highly motivated.

The one-to-many end user approach involves the traditional classroom situation—one instructor and a number of students. In computer training, other factors should be considered, specifically, student-instructor ratios, student-computer ratios, and, if tutors are used, the student-tutor ratio. This type of training may employ the buddy system, in which learning teams of two or more people per machine are formed. An important variation of the one-to-many

Fig. 5.3. Computer assistance is useful for 4-H projects, as this leader demonstrates.

Fig. 5.4. Students get computer experience in a "hands-on" workshop.

method involves defining the many as individuals who must return to their home base to train others. The actual training of trainers does not differ from the usual one-to-many model. The main distinction is that trainers must be highly motivated, since they must impart the same kind of training they are charged with learning.

Content for Training

User training should focus on the question, "What can computers do for you?" This task orientation is consistent with our interest in users rather than operators. The examples of computer-assisted tasks listed in Chapter 3 provide the trainee with relevant information on the rationale and motivation for the training efforts. Further, such a list reinforces the notion that computers are tools to use to achieve an end, not ends in themselves.

Once potential uses have been described, a brief discussion of computer terms and concepts should be conducted in sufficient depth to indicate the basic components of a computer system and the functions each component performs (Chapter 2), and the distinction between hardware and software should be made (Chapters 3 and 4). The remainder of training should cover five general classes of programs: word processing, data base management systems, electronic worksheets or spreadsheets, communications or terminal programs, and graphics packages. One approach to training in the use of generalized application software such as these can be summarized as follows: (1) Demonstrate an easy-to-understand application; (2) assign a similar, controlled application; and (3) assign a user-defined application.

Data Base Management Systems

A particularly effective way to introduce users to data base management systems (DBMS) is to apply this class of software to a task to which most users can easily relate. Choice of the example application is critical, since the emphasis is on learning to use the program and not on the nature of the task being performed. A good example is the creation and use of a mailing list for a club or organization. Ideally, the instructor identifies the task to be done and then demonstrates the creation of the electronic form to collect or record the data and subsequent use of the data to prepare mailing labels.

The user is then assigned a similar task (e.g., to create and maintain a list of 4-H volunteer leaders). After successfully completing this task with or without instructor assistance, the user is given the assignment to complete a task of his or her selection using DBMS, with or without instructor assistance.

Electronic Spreadsheets

An effective task to introduce an electronic worksheet is estimating the cost of a craft item. Such a task is easily understood and usually does not involve complications such as accounting for capital expenditures. Such an example was used by one of the authors in working with extension homemaker club members. The usual procedure is to identify the task, e.g., estimate the cost of producing a patchwork quilt for sale at a fund raiser. The instructor demonstrates how to prepare a format enumerating each cost item, quantity used, and price per unit. The cost of all items are then summed to form an estimate of the cost of producing the quilt. The worksheet thus prepared is called a template. The instructor then demonstrates the ease of making changes, such as adding an omitted cost item or updating the price for an item. The follow-up controlled assignment might be to modify the prepared template (text given user that defines values input) to reflect a different set of cost items and their prices. The final stage might be to apply the electronic spreadsheet to a task of particular interest to the user.

Word Processing Packages

Some word processing programs are difficult to teach because they use complicated procedures to make changes. Many seasoned trainers argue that rather than beginning user training with word processing, the correcting aspect of word processing programs should be introduced first. One word processing training strategy is, before instruction begins, to ask the trainee to enter a form letter on a disk, deliberately making some mistakes. The trainee then can be shown how to access the form letter and how to change the addressee's name. The typographical errors can be corrected; short phrases can be inserted into the text. A second assignment could be to make major modifications in the letter, such as adding a new paragraph. The third assignment would be for the user to start with a blank screen and carry a writing project of interest through the processes of entry, correction, and printing. A user who can complete this last stage can be advised to consult the user's manual when problems arise and is on the way to becoming an effective word processor user.

Graphics Programs

Graphics programs are a critical feature of the suggested training program because they are effective in summarizing information for both descriptive and analytical purposes. A relatively painless way to introduce users to what graphics programs can do is to prepare a bar graph showing the age distribution of users in the class. Broad age intervals (e.g., less than twenty, twenty to forty, forty and over) might be established and a quick tally taken of the class. These data (age intervals and number of users in each) can then be entered into the graphics program and displayed as a bar chart on the video display monitor. The instructor might also show how the bar graph can be printed. A follow-up assignment might be to return to the exercise of determining the cost of producing the patchwork quilt and the percentage of total cost accounted for by each cost item. The assignment is completed successfully when a user can print a bar graph and/or pie chart showing the distribution of costs. The final assignment would be one in which each user could choose a graphics problem of particular personal interest.

Terminal Communications Programs

Generalized training strategies for terminal communications programs should probably be restricted to teaching electronic mail and data base access, omitting instruction on mainframe computer capabilities. One approach to electronic mail training is first to demonstrate the procedures for signing on and reading mail for sample messages. A second assignment might require users to sign on the mail system and access messages especially prepared for each user. The final assignment might be to have each user send a mail message to another user in the class. The key to an effective introduction to accessing data bases is to identify a data base that is easily recognized by users in the class, such as the United Press International (UPI) data base on the Telenet system (see Chapter 6). Most users need little or no introduction to this data base. The initial assignment would be a demonstration of how to access UPI and print out a few news stories or view them on a monitor. A second assignment might be to search the UPI data base for all news articles on agricultural legislation during a specified time period. The final assignment might be to review all categories of news covered by UPI and to print out a story of interest.

Matching Training to User Needs

A particular training program need not cover all five classes of programs: Users only need to acquaint themselves with programs in which they have particular interest. Furthermore, users do not have to gain complete knowledge of each type of program; they can learn only as much as they need to perform the task that they wish to do. Some users estimate that they can do 95 percent of their work using these five general classes of programs.

Throughout this discussion we have not mentioned the necessity of learning to write computer programs. In the traditional sense such programming means to provide instructions to the computer written in some relatively high-level language, such as BASIC or FORTRAN. Actually the user of these five classes of programs is in fact programming by using an even higher-level language inherent in these program classes.

At least two existing or forthcoming computer programming packages integrate all five program classes. Symphony from Lotus Developing Corporation and PFS from Software Publishing Corporation are packages that not only perform each of the functions described but are written in such a way that a user can move easily from one function to another.

Implications for the Classroom

Training conducted in the classroom is within the domain of group training. Perhaps the two most critical features in effective training are highly motivated trainees and skillful instructors. In addition the following factors are crucial to effective classroom training on computer use. The first factor is availability of computer hardware: Trainees without access to computers outside the class are severely limited in their ability to use the computer to solve problems or perform tasks. Computer training without hardware is very much like teaching someone to drive without an automobile.

Second, the student-teacher, student-monitor, and student-tutor ratios are very critical in computer training. Trainees want to ask many questions, but a large class (more than 30 students per instructor) tends to preclude the question-and-answer exchange. During demonstration periods, an important limitation on class size is the ability of student users to view the video display monitors comfortably; effective instruction requires monitors in sufficient numbers and of adequate size. Related to student-teacher ratios are tutor-student ratios.

The most effective classroom training has used on-site tutors to help students during the hands-on phase. Experience in western land-grant universities suggests that an optimal tutor-to-student ratio is 1:3.

Third, many classroom training programs employ the buddy system, in which individuals are assigned to groups of two or more people and instructed to work together and help each other. Considerable controversy focuses on the effectiveness of the buddy system. Many feel that the more dominant personalities in the group will take over and the less dominant personalities will not receive hands-on training. Others feel that the buddy system is very effective because it allows peer interaction without intimidation. Fourth, the availability of well-written reference manuals and the use of examples are as important in computer training as in any other training situation.

Last, the order of presentation has been cited as critical to effective training. Many contend that training should start with familiar aspects such as word processing. Others, believing that word processing programs are complicated and should be introduced last, favor starting with electronic spreadsheets or data base management systems. The latter group introduces word processing by using the editing functions to correct items in spreadsheets and/or in a data base management system.

Two particularly challenging problems are associated with group training. Different levels of computer background of participants make the selection of the level of presentation especially difficult. If it is too advanced for the inexperienced trainee, it is too elementary for the advanced trainee. Also, in a group situation some individuals invariably like to work ahead during class. These individuals often run into problems and ask questions about material the instructor plans to cover later. This kind of disruption is practically unavoidable.

Implications for Extension Training

The implications for the classroom apply directly to extension training that occurs in groups, but three additional points are worth noting. First, in an extension context ultimate uses probably should receive more emphasis. The trainer should spend considerable time with trainees clearly showing how computers can be applied on the farm, in the home, and throughout the community. Simple examples progressing to more realistic applications are in order. Clearly related to the notion of ultimate uses are realism of the problems (tasks) addressed and trainee motivation. In recent training programs for extension specialists and county agents at one land-grant university, instructors found that motivation was paramount. Every trainee was instructed to bring a real problem to the session, but not everyone complied. Those who did were more highly motivated and better focused and as a result gained much from the training. Those who did not bring a real problem and worked only with sample problems exhibited less motivation and gained little from the training.

Second, the time period for training should be selected carefully. Twenty-four hours of training could be offered in a training program consisting of an eight-hour session on each of three consecutive days or one made up of weekly three-hour sessions over eight weeks. The latter schedule might appear easier for instructor and students, but some evidence indicates that when training programs extend over time interest drops off very quickly. Since extension clientele often cannot commit time over a long period, many trainers believe that intensive training is probably more effective.

Third, a resident expert and/or resident enthusiast must be located in each county office to provide moral as well as programmatic support to the computer applications training. Without such a person, computer applications seldom advance. Such a resident expert and/or enthusiast can be trained in group sessions at the state office or in groups of counties. Such training allows individuals to meet with peers in surrounding counties to examine common problems.

Implications for Administrators and Managers

"Real Managers Don't Use Computer Terminals," the title of an article in the *Wall Street Journal* (February 7, 1983), implies that computers are used for detailed tasks not usually performed by managers, administrators, and other professionals. This issue, of course, is hotly debated. Although some feel that administrators and managers must have a basic knowledge of computers to make decisions about their purchase, they believe that actual hands-on experience and use should be delegated to subordinates. On the other hand, others argue that personal computers can assist professional administrators and managers in their own jobs and should be directly used by them.

From our perspective, whether or not professional managers/administrators actually use a computer is not the question. They must, however, be exposed to training sessions designed to show computer problem-solving capabilities so that they can make informed decisions about resource allocation for computer equipment and training. Furthermore, without top-level management support, computer applications are not likely to be forthcoming.

The Manager-Administrator as a Trainer

The executive manager or administrator of any organization is confronted with deciding how to train people in the use of personal computers. Experience indicates that employee acceptance and application are more rapid if the manager/administrator is trained in the use of the software and participates as a trainer in introducing the software to employees. The manager should know, as a minimum, what types of software are available to help people with their everyday problems. If managers demonstrate the use of the programs in the training sessions of individuals in the next lower organizational tier, the latter may in turn continue with the next lower tier of employees, or the recently trained middle managers may provide the software demonstrations to the new group of students. This participation shows that the manager is conversant with the technology committed to the operation and applications of personal computers in the organization and is a leader in their usage. This approach also obviates the problem of the organizational commitment associated with the often heard statement, "I'll buy computers for the staff as soon as they can show me they know how to use them." The obvious incongruity of this statement is that people rarely learn how to use any equipment if they do not have access to it and are not instructed in how to use it.

The Manager/Administrator as a Trainee

Information availability, digestibility, ease of access, and potential advantages to themselves and their organization have motivated some managers to learn how to use a personal computer. Once they have gained access to information on project or funding opportunities, market conditions, or research and development activities that can help in their decision-making, they see the potential for merging local information and performing various manipulations. They soon want to know how to use spreadsheets and data base managers, along with graphics packages and word processing packages, to write reports, letters, and memos. At this point, managers are ready to be introduced to plotting packages that make overhead transparencies of graphs and other programs that make slides of text and figures for presentations in meetings and in reports.

Another effective entry point for manager training is the use of electronic mail and word processing packages. If the organization is linked through an electronic mail service to people with whom normal communication is through the postal service, managers will probably find they can handle much of their correspondence themselves through the electronic mail system by using word processing packages. The communications program on the personal computer enables them to use various on-line information services, and at this point they have integrated all the basic software into using the personal computer for communication, maintaining personnel records, budget analysis and decision-making.

Frequently the difficulty in training managers is to get them to commit their time: They do not feel the pressures that motivate secretaries, clerks, and other staff, who must grasp specific applications of software for their jobs. The motivations for managers to learn to use personal computers are productivity, competition, curiosity, and peer pressure. In training, the learning process must be made readily available and not too time consuming. However, the manager must recognize in advance that a high initial level of time commitment is essential. Once familiarity with the various software packages is gained, continued use will expand the manager's capability, and tasks that once took considerable time will be accomplished much more quickly.

Top-level executives and administrators are best trained one-on-one in their offices or separate from a group setting. In general, managers do not respond well to software simulation packages and manuals, no matter how friendly. The perceived or real conviction that they are too busy and their time is too valuable is the most serious deterrent. Trainers or educators of managers need to interact with the students, personalize the experience, and make it fun. They need to know the appropriate software that will be useful in the activities of a specific manager and of people in the organization. The ideal trainer of administrators or managers has a background in computer technology as well as knowledge of the problems, approaches, and solutions to problems faced by management.

International Training

Until the mid-1970s, the purchase of a computer was beyond the financial means of most people in the United States and in other parts of the world. Agencies and groups throughout the world can now afford computers, but they need training in how to run them and how to use the available software packages. The approaches to international training are roughly the same as those already described, but getting the trainer to the students or the students to the trainer is more difficult.

One scheme being used is the establishment of a regional training team, often composed of a core faculty from more-developed countries with local people added at the training site. The local people help maintain rapport with the students, and they know what local training approaches have proved useful in the past. Countries surrounding the host country can select and send students to the training course. Computer applications software is demonstrated on personal computers, often as applied to a regional problem; for example, a typical topic might be the use of computers in renewable natural resources. The students selected for the course would probably be natural resource managers, teachers, and researchers working for the governments of participating countries. An outside funding source, such as the United Nations, the UN Food and Agricultural Organization, or some national government, typically provides the resources to assemble the training team and students and pays for training facilities and materials. A major feature of this approach is that it allows students to study regional problems almost at their source. Another way to provide international training is to enable students to come from their own countries to a U.S. university to be trained in the use of computers and the programs applicable to their work in their own countries.

An increasing number of companies, as well as universities and some government departments, provide training in many countries just as they do in the United States. Texts and teaching materials are now available in most languages, and various translation aids and services have assisted in converting useful materials from one language to another. Application software also is being developed for use in different tongues. Stand-alone training stations composed of videotapes, training texts, training software, audio cassettes, and other items are available, but the individual human contact, still so important for effective training, is hard to package.

6 / Large Data Bases, Networks, and Service Groups

Definitions

Data Base

A large data base is a collection of text—alphanumeric characters consisting of letters, numbers, and special signs like $, %, #, or @. A data base is divided into records, the size of which is limited by the computer's internal storage buffer size. A record can be thought of as one of many similar-sized pieces of data, all of which reside in a file, which itself resides in a filing cabinet. The smaller the record, the greater the number of records that can be held in a fixed amount of memory or storage. A double-density, double-sided 5¼-inch floppy disk, for example, can hold about 500,000 characters or bytes of data.

Large data bases contain information that can be accessed often over a period of time from a number of different terminals. Arranged or sorted by keys, a large data base is best searched randomly, through key words or codes that can be unique for each record, but do not have to be. The National Pesticide Information Retrieval System (NPIRS), for example, is a large on-line data base of pesticides and their allowed applications. It is an interactive system, one that queries the user in order to permit access to the data base.

Network

A network is an intercommunicating group of terminals and/or computer systems. The simplest networks are point-to-point and multipoint networks; each point is called a node and is connected, at most, to two other nodes. In a ring network, the points are connected in a circular or ringlike arrangement. A break at any point in the ring leaves an alternative pathway for data, back around the ring. A star network, which connects several peripheral computer systems to one central system, is the kind found on most land-grant university campuses, with the university's mainframe computer serving as the central system. The most complex network is one in which each point is connected to every other point. These are called fully connected networks.

Service Bureau

A service bureau is an organization that puts together large or small data bases for access by communicating microcomputers. It can provide information on numerous subjects such as the weather, crop marketing, and agricultural outlook. Charges are usually based on how long the user's microcomputer is hooked up to the service bureau's computer (connect time). Other possible charges include an initial fee for setting up the user's account and a minimum monthly fee. The growth of service bureaus presages the day when the home or farm will be an information center, able to obtain and retrieve advice and information on a wide variety of topics.

Microcomputers as Communications Devices

Many computer experts say that the future of computer technology lies in large networks of interconnected computers that share data stored in and distributed from large data banks (see Chapter 1). Anyone with a computer potentially can communicate with anyone else with a computer. Networks of computers that encompass county extension service offices already exist in some states and are useful for a variety of collaborative undertakings. Possibilities for communication range from the simple dumb terminal with modem hookup to the microcomputer serving as an intelligent terminal, able to store and access data.

Hardware Requirements

Computers can be connected into a network either by cables or by telephone lines. Stringing cables through walls and ceilings is quite adequate within one building; however, most transmission breaks down after a distance of more than 2,000 feet. Communicating with a computer at a greater distance can be done through the telephone wiring system.

To use the telephone to transmit data, the computer needs an additional piece of equipment, such as an acoustic coupler. Because it changes sequential trains of pulses into sounds at a given frequency, the acoustic coupler can be frustrating to use on a heavily loaded telephone system or in situations where another person can pick up an extension. A typical acoustic coupler costs from $200 to $400 and transmits data at the rate of 30 characters per second (cps). Commercial devices that transmit at 120 or 240 cps are available, but they are more expensive, and many rural telephone lines cannot handle their speed.

A modem (modulator/demodulator; also called a dataset) performs the same function as the acoustic coupler except that it connects electrically instead of acoustically to the phone line, skipping the step of going through the microphone and speaker in the telephone handset. Modems can achieve speeds of 240 to 480 cps; even speeds of 960 cps are possible when private, secure lines are used. The modem is more expensive and less portable than the acoustic coupler, but it does provide higher speed and greater reliability of data transmission.

Software Requirements

Most time-sharing systems are very forgiving; that is, they accept a wide variety of signals from a microcomputer or terminal. On some, however, the user must be concerned with technical communication procedures that may differ from network to network and are usually resolved by setting the appropriate switches on the computer terminal.

Sign-on procedures differ widely from network to network and from service bureau to service bureau. To keep records for billing purposes and to prohibit unauthorized use of data, most sign-on procedures require users to specify an account number, access code, and confidential password; sometimes all three are needed. A typical sign-on procedure for the ITT Dialcom network used by the USDA Extension Service involves the following steps.

User Procedure	*System Response*
Dial or autodial access phone number.	High-pitched tone sounds.
Hook phone to acoustic coupler or modem.	Connect light goes on.
Hit return key.	PLEASE SIGN ON appears.
Type in logon ID and password.	Items or notices of interest, then > (prompt) are shown.

Issue commands to access ITT
Dialcom features, such as
MAIL READ or MAIL
SEND.

Telephone networks have been created to minimize long-distance charges when the user is not located within the local call area of the host system. With ITT Dialcom, for example, GTE Telenet and Tymnet services are available to almost all users. Access is made through a nearby node via a local phone call.

Specific Applications

Market News (Service Bureau). Agri-Data is a private information computer network that specializes in agriculture and offers access to more than 10,000 pages of marketing, business, financial, weather, and price information. It continually updates its data and offers several services, including an on-line computing system that allows users to transfer a library of microcomputer software to their own microcomputers.

Agriculture Marketing Service (AMS) market news network reports up-to-date information on commodity prices, demand, and movement. Data are transmitted by the USDA Agricultural Marketing Service to the news media and among market news offices. Firms and individuals may also subscribe at their own cost.

Pesticide Data (Large Data Base). The National Pesticide Information Retrieval System (NPIRS) is a nationally accessible on-line data base that contains information about all pesticides registered with the Environmental Protection Agency. NPIRS indicates which pesticides are registered for use against specific pests on specific crops or sites.

Software Reference

The most comprehensive software guide for the Cooperative Extension Service was published in early 1982 and is available from the University of Florida Cooperative Extension Service. Under a cooperative agreement with USDA and Virginia Polytechnic Institute and State University, this inventory is being updated; the update is available both in printed form and through the Computerized Management Network (CMN), which has users in the United States and Canada. Various private directories of software are also available for particular brands of hardware. The user can get these directories at computer stores or through mail-order sources. Trade journals frequently publish tables of software packages for a specific use, comparing them and noting hardware and operating system restrictions.

Electronic Bulletin Boards

Electronic bulletin boards systems (BBS) are being used more and more by computer owners with a communications capability to share information. Bulletin board users call up a host computer to retrieve and store data. If the host machine is a personal computer that needs to store a large amount of data, a hard disk is generally used because a floppy disk would be quickly filled with only a few BBS users. In many ways an electronic bulletin board works just like a conventional one: Both allow a user to post and read messages, addressed either to everyone or to specific groups or specific individuals; both may be used to notify others of items wanted to buy or sell.

The major difference is that a BBS may also provide listings of programs available to be transferred to a personal computer (downloaded) and later run on it. In addition, program listings may be transferred from a personal computer to the BBS (uploaded) so that others may later download and use them. Bulletin boards vary considerably in their use, contents, and sophistication, but most address specific topics or provide certain services. For example, one may be oriented toward messages whereas others may address such areas as dating matchups, movie reviews, or program listings for specific brands of computers.

Some bulletin boards operate on a personal computer with only one phone line and heavy competition for use. This situation may lead to limiting the number of acceptable users and assigning names and passwords, as well as limiting the total time an individual can access the BBS each day. Certain BBSs charge for their use.

Many computer user groups now share information through a BBS with members using either their computers with modems to communicate with the BBS or dumb terminals. Most BBS callers using a personal computer employ some form of communications software such as Crosstalk or Modem-7 (the latter is in the public domain). Several communications programs also allow the personal computer to be used as a bulletin board. A major feature to look for is telephone answering on the part of the modem. Many versions of bulletin board software are available in the public domain and are accessible on many bulletin boards. Names, general coverage areas, and telephone numbers of BBS services or operators may be found in computer stores or in books, magazines, the Computer Shopper, or from user groups. A partial list of available data bases, networks, and service groups is presented in Appendix D.

Regional Computer Applications Centers

The United States has been divided by the federal land-grant university system into four regional groups of land-grant university states based on their similar climate, agriculture, and problems: the northeastern, the north central, the southern, and the western. W. K. Kellogg Foundation has entertained proposals from each of these regions for the establishment of regional computer applications centers or institutes designed to assist land-grant universities in sharing computer information among themselves and with state Cooperative Extension Services. The extension services can then transfer such information to ranchers, farmers, home owners, and small business operators—in general, to their state's clientele. The Kellogg Foundation has joined with states in three of these regions to establish computer institutes: (1) the North Central Computer Institute, hosted by the University of Wisconsin, Madison; (2) the North East Computer Institute, hosted by Pennsylvania State University, State College; and (3) the Western Computer Consortium, hosted by the University of Arizona. A feasibility study was recently finished in the southern region, and that area is now ready to submit a proposal to Kellogg to establish a similar center there.

Several studies of current computer-related activities in the north central and northeastern land-grant universities have indicated that some form of regional assistance might be beneficial in the transfer of computer information from the universities to individuals using the technology. Because of lower prices many persons in small businesses, farms, ranches, and homes can now buy personal computer hardware and software for use in their day-to-day endeavors. Students in elementary and high school classes are exposed almost daily to the computer as a tool for study and entertainment. Soon there will be millions of these devices located throughout society. At the same time, public demand for computer education and guidance has greatly increased, with institutions being severely challenged to maintain their historical leadership role.

On the campus computer resources are being stretched to capacity, and technical support staffs are being asked to respond to more requests for assistance than they can physically handle. Research scientists are using programmers and students to write software and gather data for analysis and study, adding more competitors for the already overloaded computer facilities. Cooperative Extension Service specialists and agents have found that computers facilitate the delivery of information to their clientele and speed up their routine data-handling activities, thus giving them more time to use in their programs. They, therefore, are now competing for the same scarce computer resources used by research scientists and students.

The three regional computer application centers or institutes already established have resulted in savings for the land-grant universities and their outreach organizations such as the Cooperative Extension Service. The regional centers are designed to support the extension, research, and instruction functions in agriculture and home economics. Membership in the regional centers is open to all interested land-grant universities within a region. Each center serves as a vehicle for information exchange among its members and assists them in their programs of computer-related information transfer to extension clients. Although the objectives of the centers differ somewhat from region to region, they share several common goals:

- Facilitating the development of multidisciplinary computer software.
- Advancing the interinstitutional sharing of computer software and other information.
- Developing and sharing regional data bases.
- Sponsoring the use of educational services, techniques, and products to train individuals in research, extension, and teaching at member institutions.
- Providing assistance in contacting individuals and groups knowledgeable in computer products, services, and applications.

Forms of Organization

The regional computer centers have slightly different organizational forms. Both the North Central and North East Computer Institutes are central facilities with a number of staff members who provide services and information to the member states. *States* in this context are those land-grant university elements for a given state in the region that wish to be associated with the institute and help bear the financial burden of its operation. The staff of the Western Computer Consortium comprises a director, a secretary, and a part-time editor. The rest of its personnel are faculty/staff members at the land-grant universities. In the west, the feasibility analysis revealed that most states had one or more areas of excellence in computer applications. A portion of each state's payment for maintaining the consortium is some fraction of the time and activities of an individual knowledgeable in that area. This arrangement allows each member institution to share its expertise with other consortium members. For example, one state might be strong in maintaining publications inventories whereas another might be strong in computer-assisted soils analysis and recommendations reporting.

Each of the three funded centers has a board of directors comprising appropriate representatives of the directors of extension, research, and instruction at the member land-grant universities. The board of directors sets policy and sees that it is followed. Board members are assisted by a technical advisory committee composed of one computer-knowledgeable individual from each member institution. A campus coordinator is also selected at each member institution to assist in the everyday information flow from and to the regional center and the member states.

Functions

Reduction in Software Duplication. Perhaps the most important objective of the regional centers is their attempt to reduce, if not eliminate, duplication of software development in the region. They try to coordinate the design and development of regionally applicable software so that member states may share in development activity and use the final product. The centers also facilitate the sharing of each state's experiences in software/hardware purchase, evaluation, and use, as well as detailed information on specific questions. Member states may use and build on the other states' experiences; they need not repeat a bad experience.

Linkages. The regional computer application centers are linked to each of their member states through the board of directors, technical advisory group, and campus coordinators. They are also loosely linked to each other to maintain an awareness of each center's activities and potential for sharing. They utilize several on-line information services and pass some information from these through the centers to the states. In some cases, the states then pass it on to the extension clientele. The centers are formally or informally linked to agricultural extension, research, and instruction performed at the land-grant universities.

Information Sharing. What does all this mean to an individual whose work is related to agriculture? For one thing, the information and services distributed by these centers may be directly useful. The North Central Computer Institute has established an electronic bulletin board to transfer agriculture-related information and microcomputer programs from the institute to member institutions for their distribution to the farmer, urban dweller, agricultural researcher, and others. The bulletin board also allows messages to be sent and received from other users of the BBS. Microcomputer programs may also be sent to the center on the bulletin board for later downloading and use by other users of the system. Software reviews and other computer information are available on the bulletin board and free for the taking. This bulletin board asks that only persons interested in agriculture use it.

Publications. The regional centers also publish various items that may be used by the member states. A typical publication is only a few pages long and deals with a specific topic such as software design or descriptions of videotext. Most centers also publish some form of newsletter covering items of both regional and general interest. The nearest regional center should be able to provide a list of available publications.

Addresses

North Central Computer Institute, 610 Walnut Street, Madison, WI 53706. Tel: (608)263-4791

North East Computer Institute, 1315 South Allen Street, State College, PA 16801. Tel: (814)863-4678

Western Computer Consortium, University of Arizona, Tucson, AZ 85721. Tel: (602)621-4330

7 / Major and Emerging Issues

This chapter presents a number of new, emerging, and generally important issues that the user, buyer, and owner of a personal computer should know about.

Videotext

Many different versions of systems go by the name videotext. The most common system is one in which direct two-way communication is established between an individual and a host computer. A typical installation is a standard television set connected to a teletext device with a keyboard; the teletext device is also connected to the local telephone service. The individual calls the videotext number and observes menus on the television screen, showing what information is available. Just as a person using a personal computer with a modem accesses an information source or service, the teletext user selects the information desired, possibly the sports section of the current newspaper, and keys in this request to the teletext device. The request is subsequently sent over the telephone line to the computer, and the requested information is returned over the phone line and displayed on the user's television screen. In some systems the requested information is received and stored in the local device and the phone connection broken. The individual may then read the received information at leisure, minimizing telephone connection time and charges. In other systems the telephone remains connected for the duration of use.

The types of information available vary with the region of the nation; some localities have versions of videotext whereas other do not. Several large companies, entering this type of system technology on a national scale, offer to the individual fairly comprehensive services such as videotext, information retrieval from various data bases, electronic mail, at-home shopping, at-home banking, remote computer services and use of programs, downloading of programs, news and weather reports, stock and commodity information, and so on. Both videotext and on-line information services are dynamic areas with high-growth potentials.

Data Base Ownership

More and more data bases are being made available to the on-line user. In general, these data bases are developed and loaded by a company that then sells use of the data base. The data base user makes queries to the data base and receives the appropriate information. A question arises as to ownership of data when the user of the data base wants to market retrieved data. If the data base user retrieves a large portion of the main data base, may that user then turn around and sell it, or access to it, in competition with the original developer of the data base? Ownership of data is currently not clearly defined nor are there rules governing how data can be used. Clarification of these important issues is expected soon. Meanwhile, several companies offer for sale or access a variety of secondary data bases.

Operating Systems

Many operating systems have been developed for the multitude of personal computers that exist today. The early standard operating system, CP/M (which exists in 8-bit and 16-bit versions) now has a major rival in MS-DOS (which has only a 16-bit version). Although MS-DOS may be the leader in the marketplace for the next few years, it in turn has a serious competitor, UNIX. The UNIX operating system differs from CP/M and MS-DOS in several ways: It is designed for multiuser, multitasking where each user has full access to system resources according to scheduling and to memory swapping algorithms not found in the design of most of its competitors; and it is transportable, because it is written in C, a machine-independent language.

Another competitor is the Pick operating system, which also supports multiusers and multitasking operations. Its major feature is its ability to perform data management easily. In 1984, Pick will release a newer version of its operating system that will give it the ability to run UNIX, CP/M, and MS-DOS in emulation (that is, to use the commands of these operating systems as if it were using the systems themselves). It will be able to operate in a multimachine environment as a distributed data base; such a data base is divided among several computers or in several user areas. Finally, applications may be written for it more easily because it will have a more comprehensive command set.

Anyone buying a new computer should look carefully at its ability to support several operating systems or use software developed for several operating systems such as CP/M, MS-DOS, and UNIX. In deciding between a computer that uses CP/M and one with MS-DOS, cost is a major consideration: A CP/M machine costs about one-half as much as a MS-DOS machine.

A 16-bit machine has more memory, as well as potentially faster computational speed. However, not all 16-bit software is necessarily more powerful than 8-bit software. A number of 16-bit applications programs were developed for 8-bit computers and then translated, line for line, to arrive at a 16-bit program. They have not taken advantage of the more powerful 16-bit computer processor chip. In some cases the computer may utilize the 16-bit software but only pass eight bits of data on the bus, or information path, at one time (the same as an 8-bit machine), a technique that reduces the 16-bit efficiency. The IBM PC is an example of this.

A large inventory of computer programs has been written for 8-bit machines, which addresses an enormous range of areas and applications. At the same time, some excellent applications programs are available only on the IBM PC and compatibles. The 16-bit machine also may be able to support the UNIX operating system, a possibility which might sway the buyer toward that machine. Whatever the choice, the user is likely to find software to satisfy primary needs. Computers can be compared with cars—they will all get you from A to B, but personal taste and need primarily will dictate what model someone will buy.

Software Issues

Software Prices

Several nationwide surveys have found that computer software prices will be forced down by market competition for the next few years. The prediction is that during the 1984-1987 period the price of a disk-based software package for home use will drop. This is a reversal of the price increase from 1982 to 1984 caused by the growing complexity of software, such as spreadsheets, that was moving into the home. The same surveys showed that entertainment software was the most popular kind bought for the home (36 percent of all packages); next was educational software (22 percent of the total). Entertainment software is expected to be the fastest growing category in the near future and to experience the sharpest price drop. The user should shop around for software because prices differ greatly.

Configuration Management and Software Service

The matter of configuration management needs to be considered by both the developer and the buyer of software. As the software is used, possible problems are found that may require attention and modification of the code. These modified versions of the code, sometimes called fixes, should be distributed to all individuals who have bought the program and are using it. Finding problems, modifying the code, and transferring the improved code to the user is called configuration management. It can be a costly feature for the developer, and the buyer should ask if this service is part of the software purchase. Again, it is worthwhile to shop around. Some software developers and sellers disavow any knowledge of the customer once the sale is made. Others help the user learn the operation of the software, show how it can be used for specific applications, and provide the latest fixes or revisions. The best arrangement is the one in which both the software developer and the seller provide assistance in the operation, fixes, and enhancements to the software after purchase.

Piracy

A major problem in the software industry is piracy or unauthorized duplication and distribution of software. There is a growing interest in having national laws passed to allow the legal system to punish pirates. Louisiana became the first state to introduce such legislation, with its Software License Enforcement Act, which asserts that computer software is protected under U.S. copyright laws. This act gives the copyright owner, in this case the software developer, exclusive rights to control copying, distribution, and modification of the copyrighted item. This legislation defines the software user as merely the licensee of the software. Title to the software is retained by the licensor—the software developer. The law also requires clear labeling on all software packages, using large lettering to convey the limited rights of the end user. One estimate is that more than $360 million was lost during 1983 through software piracy, and an amount near $500 million will be lost in 1984; another estimate places the loss to software publishers in the billions.

Locked Software

Several companies have entered the software protection business. One of the most prominent of these, Vault Corporation, sells publishers diskettes altered by the proprietary Pro-Lock technique. This alteration physically identifies the diskette, just as a human fingerprint indentifies its owner. Software can run on the disk only if it is encoded like the diskette originally packaged by the software publisher. Another system, developed by Ashton-Tate, employs electromagnetic signatures that can be read but not written, software concealment of major features, control of the duplication equipment, and frequent changing of protection variables. No protection system has remained uncracked by enterprising programmers for more than a few months. However, the devices do remain effective against casual corporate and home users.

Language Translators

The increased emphasis on the C computer language and the UNIX operating system does not mean that the user must sit down and learn how to program in C. Several language translator programs now available take specific forms of BASIC-coded programs and translate them into computer code in the C language. Since a number of different versions of BASIC are in use, each with a slightly different instructional set, no universal translator package can accept all versions of BASIC. Each form or group of several very closely related forms has its own translator; this arrangement allows the enormous number of existing BASIC programs and algorithms to be transferred and used in the C language. Other language translators also are available. They can usually be accessed through a university or large computer center. The cost of the BASIC-to-C translator is roughly $400; other translators are usually much more expensive.

Software Validity

Some application software programs are written and tested for specific geographic regions. When bought in one region and used in another, they may yield incorrect predictions. An irrigation scheduling program can easily be geographic specific. For instance, a program designed for scheduling the irrigation of wheat in Montana might not work well for cotton crops in Arizona because of the major variations in climate, soils, and types of crops between the two states. The range of temperatures during the growing season in Montana might be outside the range of growing temperatures in Arizona. The use of temperature data outside the range of those used to develop the model can easily result in erroneous output or predictions. The user needs to be careful in using software, making certain the documentation clearly defines the conditions or regions where the program is valid. The same is true of the range of variables that may be used as input to the program. The user may exceed the acceptable input range and obtain predictions that are not valid or logical: Just because a program outputs numbers does not mean those numbers are appropriate. Anyone in doubt about the ranges in data or conditions over which the program is to be used should ask the software developer if possible. For an agricultural program, a local Cooperative Extension Service agent may be able to help.

To avoid legal liability, most programs that might be used in areas for which they are not designed or those that are used to make economic decisions are marked to indicate that the user bears all liability in the use of the program. The developer and seller disclaim any liability in the consequences of its use or misuse. In many cases, modifying or changing the parameters and coefficients used in programs designed for one area will make the predictions valid for another specific area. For many programs, the appropriate numbers are known by researchers and extension personnel in the land-grant universities. Before buying a specific predictive model it is a good idea to see if it has been used successfully locally.

Integrated Software

Some of the best-selling software products are what have been referred to as "integrated" software. These packages provide the user with the ability to exchange data between applications without having to return to the operating system to change programs. The user also does not have to be concerned about compatability between applications and data formats. Almost every major software developer either has marketed or is just putting the finishing touches on integrated operating environment software that features on-screen windows. These packages, for 16-bit machines, make it easier to get full productivity from a personal computer.

Windows are simply separate areas or boxes on the computer display screen. They are especially useful for systems integrating several applications, because each window can display a different aspect of work. For example, in preparing a textual report that will include graphs of spreadsheet data, the user can display a word processing program in one window, a graphics program in another, and a spreadsheet in a third. Although a user still has to switch modes (change the main program used in the foregound) when moving between applications, in most current packages the windowing system makes it easier to refer to the spreadsheet while writing the report, to create graphs by transferring figures from the spreadsheet to the graphics program, and to move finished graphs into the appropriate spots in the word processing document. Some systems permit performing such tasks with a mouse (a desk top device, which when moved by the user moves the computer screen's cursor) rather than keyboard commands. Most systems overlay the computer's operating system and present a much friendlier and more consistent interface with the machine. Major new applications written specifically for the windowing environment probably will not be available until early 1985. It might be prudent, before buying, to see how the various systems compare.

Friendly Data Base Management Systems

One of the major selling features of data base management systems (DBMS) has been that they are relational; that is, they can be organized by terms that relate to specific fields of data. For instance, a field containing individual names could be called NAME, the user then could sort names by issuing the command SORT NAME/A. The current trend relates DBMS to people. The DBMS now being developed go beyond being user friendly to being able to be used by the computer novice. Many software companies have information management packages or applications written in a data base language. In some cases data entry screens can be designed to look exactly like existing manual record forms, thus easing the transition to computerized records management. Other DBMS guide the user through the process of setting up a data base by menus that can be skipped as the user becomes more proficient. In some new programs the integrated software is built around a DBMS. The nature and characteristics of DBMS are currently changing, whereas older systems such as dBase II are being changed to make them upwardly compatible.

Mainframe-Microcomputer Communications

An ever-increasing number of computers and information services are becoming available to the personal computer user who has a modem or other means to access the public telephone network. One group providing information, computer usage, electronic mail, shopping, stock market information, and numerous other services is The Source. Either a terminal or personal computer may be used to access and utilize its services. Some form of communications software also allows the personal computer to communicate with the distant mainframe computer.

Another means of linking a personal computer and a distant computer is through an intelligent link, a communications link in which both the personal computer and the distant computer run a given software package, such as a spreadsheet. The intelligent communications link allows data to be requested and moved through the link from one program to the other without leaving the user level of the applications program. The major problems are cost and the scarcity of programs. The personal computer software to allow linkage for a specific applications package might cost $300, whereas the price of the package for the large computer might be in the range of $30,000.

Hardware Issues

Mass Storage

Mass storage of information is one of the most critical elements of any computer system; some place must be available to store information when the computer is turned off or another program is being used in the memory. The speed at which stored data moves back and forth between the computer and a mass-storage system is a key factor in determining overall productivity. Emerging technologies that promise more and faster data storage in the future include perpendicular magnetic recording on both hard and floppy disks, laser disks, and laser cards.

The floppy disk, the most popular medium for mass storage on personal computers, is being augmented by some users with the hard disk, which has larger and faster data storage. Hard disks may be 3½, 5¼, 8, or 14 inches in diameter and are either fixed or removable. The best sellers are removable 5¼-inch cartridges that store 5 megabytes or more on one side (1 megabyte equals 1 million bytes), with up to a maximum of 20 megabytes using both sides. Many hard disk drives fit into the space once occupied by the floppy disk drive. Adding the components to use a hard disk to an existing computer can be very complicated. If possible, a dealer should install the disk and check it out to make sure it is compatible with the user hardware and software. Buying a computer with a hard disk already built in by the manufacturer is probably best.

A videodisk or laser disk can store up to 54,000 frames; that is about 30 minutes of motion at normal speed on a video player (assuming the frames are part of a movie). Some forms of video players allow specific frames to be requested and displayed on the monitor screen. Linking the personal computer to video players with such capability and using instructional material recorded on the disk open up new areas for education and information transfer. Only selected frames in a selected sequence are presented to the user. Some frames may be text whereas others are images. In some cases, groups of frames can be displayed to show motion. The lowest cost for a video player that will interface easily with a personal computer through an RS-232 line is approximately $800.

Laser players use two types of disks. One is a constant angular velocity (CAV) disk in which a frame of video is displayed from a single rotation of the disk. This form of disk allows relatively easy presentation of freeze frames (i.e., reading and presenting the same frame over and over). Another form, the constant linear velocity (CLV) disk, does not turn at the same speed all the time, and each rotation may not necessarily contain a single frame. It does not have single frame display, but it holds twice the information of the CAV disk.

Most feature movie disks are CLV whereas most disks made for interactive use are CAV. Both types of players may be controlled by a personal computer. The CAV disk has greater potential because it can have single-frame presentation. Applications currently being developed that combine video-recorded data and the personal computer range from games to computer-assisted instruction packages. There is enormous potential for tying together video and computer technologies.

Used Computers

As new machines with more and better features become available, some people want to get rid of a computer and purchase a new one. A potential buyer or seller of a used computer, faced with determining a fair price for a specific model, can get help from the *Computer Classified Bluebook:* Like that for used cars, the computer bluebook lists reasonable prices for used personal computers. However, this computer bluebook, published quarterly, costs $85 a year, compared to $3 for the bluebook for used cars. A library or a computer store may have a copy.

Keeping Up with Technology's Onslaught

A good source of information on the stability and reputation of individual companies can be found in several weekly microcomputer news magazines, such as *Infoworld, Computer Retail News, and Computers + Software News.* The latest news and gossip about such subjects as Chapter 11 bankruptcy proceedings are reported there. *Infoworld* and other magazines provide some of the least biased, albeit controversial, reports in their weekly reviews and cumulative review capsules on new hardware and software.

All magazines have admitted or unadmitted bias. A magazine devoted to a particular computer model can certainly be counted on to tout the particular strengths of that computer. However, one of the first things a user of a new computer should do is subscribe to a magazine *devoted to that computer.* Because technology is moving so swiftly, this subscription is the best way to be advised about new products for a specific computer. New manufacturers cannot afford to advertise in general purpose journals; they must target their advertising to specific machines. The following is a list of some of the magazines and their associated computers (see Appendix C for a larger periodical list).

Microcomputer	*Journal*
IBM PC	*PC*
	PC-World
	Softalk PC
Apple	*CALL A.P.P.L.E.*
	inCider
	Nibble
	Softalk
Osborne	*FOG Newsletter*
	Portable Companion
TRS-80	*Microcomputing*
TRS-80-Color	*Hot CoCo*
Software helps (dBase II)	*DataBased Advisor*

Many of these magazines are published commercially and supported by advertisements; others are published by user groups and supported entirely or partially by user fees. Joining a national user group can be particularly useful if the computer company goes out of operation. Members of FOG (First Osborne Group) have been kept apprised of every development in the reorganization of the Chapter

11 proceedings of the Osborne Corporation. Members of many user groups such as CALL A.P.P.L.E. (Apple Puget Sound Program Library Exchange) have access to many public-domain programs at very low prices. Most important, members often exchange information on little quirks or specific uses of systems that most manufacturers or distributors cannot afford to provide.

A quick perusal of any microcomputer magazine reveals a wealth of computer hardware and software discount firms. Although well-established, distributor-supported software can be an excellent buy, hardware purchased at a distance can be a disaster if equipment problems develop. However, if the user is well versed in hardware maintenance electronics, and is willing to repair and maintain the equipment with only sparse manufacturer support, a great deal of money can be saved by purchasing from mail order or discount houses. Those who use this strategy should have a strong background in electronics and a great deal of patience.

International Agricultural Applications

The USDA Extension Service addresses computer problems similar to those found throughout the world. As a consequence, computer software that assists extension clientele in the United States can be modified to address similar problems in other parts of the world; indeed, such modifications are being made by various universities, consulting groups, commercial companies, and U.S. and foreign government agencies. Software developers are in short supply in many developing countries; consequently, there is a demand for agricultural and related resources applications software from the United States and similar countries.

For a while, the use of personal computers in foreign countries was rather risky as service was often far away and long in coming. This situation is changing rapidly so that service is often relatively close by and timely and parts, new computers, and peripherals are available in most major cities in the world. Still, the computer brands being sold and the service and parts available differ from country to country and among various parts of the world.

Several problems may have to be overcome to transfer computer applications programs from the United States to another country. The documentation for the program may have to be translated into the language of the country or one in which the user is fluent. Coupled with this is changing the program text presented to the user so that it is in the same language as the documentation. If the applications program is a modeling or prediction program, it needs to be examined to see if it is valid for the conditions found in the country. If not, the relationships and coefficients need to be adjusted, if possible, to yield valid predictions. These adjustments must be educated guesses at times as the data available on the specific parameters necessary for calibration are often limited.

As the Cooperative Extension Service provides assistance to growers and consumers in the United States, so do the universities, various government agencies, and groups similar to extension in other countries. Sometimes this assistance is from UN agencies or from governments other than the local one (e.g., U.S. Agency for International Development). Many commercial groups, ranging from computer manufacturers to training and consulting groups, are also becoming involved. Often the technology is transferred from the developed country to the local government, which then transfers it on to the producer and consumer. (For a description of computer technology transfer by regional training teams, see Chapter 5.)

Appendix A: Glossary of Computer Terms

acoustic coupler (acoustically coupled modem): A device used for computer communication over a phone line. This connecting device sends and receives computer signals directly through the mouthpiece and earpiece of the phone.

address: A number specifying a particular location in the computer's memory.

ADP: Automatic data processing.

AI: Artificial intelligence.

algorithm: A step-by-step procedure, often expressed in mathematical terms, for solving a problem or obtaining a particular result.

alphanumeric characters: Characters represented either alphabetically or numerically or using other print characters (for example, A B C D E F 1 2 3 4 5 ? $ * + −).

ALU: arithmetic and logic unit.

American Standard Code for Information Interchange (ASCII): A 7-bit (or 8-bit compatible) standard code adopted for transferring data among various types of data processing and data communications equipment.

analog-to-digital converter: A device for converting voltage levels (analog signals) to digital (computer) information.

application: The use of a computer system to accomplish a specific goal.

applications software: Programs designed to instruct the computer to perform real-life tasks (see **software**).

arithmetic/logic unit (ALU): The element of the computer that performs the basic data manipulations in the central processor. It can usually add, subtract, complement, negate, AND, OR, rotate, and compare.

array: A set of numbers or other entities specifically ordered. The elements of an array can be referred to by their position in the set. These arrays are indicated in many computer languages by subscripted variables, such as A(X), where X is the subscript.

artificial intelligence (AI): A branch of computer science dealing with the development of machines capable of carrying out functions normally associated with human intelligence such as learning, reasoning, self-correction, and adaptation.

ASCII: American Standard Code for Information Interchange.

assembly (assembler) language: A source language that includes symbolic machine language statements in which there is one-to-one correspondence with the instruction formats and data formats of the computer.

asynchronous communication: A form of serial communication not requiring common clock pulses between transmitter and receiver. Start and stop bits are used to define a character. Asynchronous means "without regular time relationship."

author (authoring) languages: High-level languages that allow the user to program without having much knowledge of a computer language. Some author languages (e.g., PILOT) determine programming needs through the user's responses to a series of questions and then provide an appropriate formatted program.

backup copy: A copy of a file or data set made for reference or safekeeping should the original file be damaged or lost.

base: The number on which a given numbering system is built. The decimal number system uses base 10, and the binary system uses base 2. Many microcomputers use base 16 (hexadecimal).

BASIC (beginner's all-purpose symbolic instruction code): A relatively easy-to-use computer language that comes with most small and personal computer systems.

batch processing: Collecting items to be processed into one file or group and then submitting them to the computer as a whole. It is contrasted with interactive processing.

baud: A unit used to measure the speed of data transmission. The term is often used as a synonym for bits per second.

baud rate: The speed at which information is exchanged over communications lines, generally expressed in characters per second. Three hundred baud is the most common rate; it is roughly equivalent to thirty characters per second.

BBS: Bulletin board system.

BCD: Binary-coded decimal.

BDOS: (basic disk operating system): The major portion of CP/M operating system. It keeps track of where to read or write information on the disks as well as handle commands between the user and BIOS.

bidirectional: A term used to depict the way a printer operates. Many smaller printers are bidirectional; that is, they print from left to right then right to left, speeding up the actual printing. In communications it refers to information moving both ways over a path.

Binary-Coded Decimal (BCD): Groups of binary digits used to represent decimal numbers, where each digit position of the decimal number is represented by four binary bits.

binary system: A two-digit numbering system based on the digits 0 and 1. It is the basis for calculations on all computers and the basis for storing and retrieving information, including alphabet characters.

BIOS (basic input/output system): Part of the CP/M operating system that manages peripherals.

bit: The smallest unit of information the computer recognizes. A bit is represented by the presence or the absence of an electronic pulse, 0 or 1.

bit map graphics: A technology that allows control of individual pixels on a display screen to produce graphic elements of superior resolution, permitting accurate reproduction of arcs, circles, sine waves, or other curved images that block addressing technology cannot accurately display.

boot: To load and run the disk operation system or some other externally stored program into the internal memory of the computer.

break: To interrupt a computation or program and return the computer control to a user.

bug: A fault or error in a computer program.

bulletin board: A computer program designed to allow a host computer to store messages, program listings, and text that are all made available to distant computers as they call in and connect over a telephone line.

bus or buss: A set of wires and connections used to transfer information between various computer components: central processing unit, input/output ports, terminals, and interfaces.

byte: A byte is composed of several bits. Eight bits are typically used to represent one character—such as a letter, number, or punctuation mark. The older microcomputer systems used eight bits per byte; the newer ones are based on sixteen or thirty-two bits per byte.

CAI: Computer-assisted instruction.

card: Printed-circuit board. Also refers to punched card.

card reader: A peripheral device that reads punched cards or forms as an initial step in computer processing.

catalog: See **directory.**

cathode ray tube (CRT): A television-like display screen that uses an electron beam to produce readable characters or graphic information. It is also known as a monitor.

central processing unit (CPU): The computer module in charge of fetching, decoding, and executing instructions. It contains a control unit, an ALU, and other related facilities such as registers, clocks, or drivers.

character set: A defined set of printing and nonprinting characters. Members of the set may vary from computer to computer. See **ASCII.**

chip: A thin silicon wafer on which electronic components are deposited lithographically in the form of integrated circuits.

CMI: Computer-managed instruction.

COBOL (common business-oriented language): A high-level programming language widely used in business applications.

code: A synonym for a computer program—therefore, a programmer-generated code.

COM: computer output microfilm.

command: A direct order to the processor to perform an operation with the information in external or internal memory.

communications network: A configuration formed when several individual computers are connected so that files or messages can be sent back and forth between large information systems and individual users.

compiler: A program that converts one computer language into another to store it for later use. It usually refers to a program that translates a higher-level language into a computer's machine language.

computer: An electronic device that manipulates symbolic information according to a list of precise (and limited) instructions called a program.

computer-assisted instruction (CAI): An instruction method for providing the student with programmed sequences of instruction under computer control. The manner of sequencing and progressing through the materials permits students to progress at their own rate.

computer, general purpose: A computer designed to solve a large variety of problems; a stored computer program that may be adapted to any of a very large class of applications.

computer language: An artificial language designed to allow communications between human beings and computer systems.

computer literacy: The general range of skills and understanding needed to function effectively in a society increasingly more dependent on computer and information technology.

computer-managed instruction (CMI): An application of computers to instruction in which the computer is used as a record keeper, manager, and/or prescriber of instruction.

computer network: Two or more computers connected so they can exchange information.

computer program: A collection of instructions that together direct the computer to perform a particular function.

computer system: The main parts of a computer: input/output system moves the information, central processing unit manipulates the data, memory holds data and instructions in temporary or permanent form, and control unit masterminds the previous three.

connect time: In time-sharing, the length of time the user is connected to the computer, that is, the duration of the telephone connection.

configuration management: The task of accounting for, controlling, and reporting the planned and actual design of a product throughout its production and operational life.

control (CTRL) key: A nonprinting character on a computer keyboard used in conjunction with other keyboard keys to send special characters to the CPU.

courseware: Computer programs used for instruction, along with manuals, workbooks, and other supporting materials.

CP/M (control program for microcomputers): A popular operating system for 8080, Z80, and 8085-based microcomputers created by Digital Research.

cps (characters per second): The rate of data transmission measured in characters transferred per second.

CPU: central processing unit.

CRT: cathode ray tube.

cursor: The indicator of position seen on a video display screen. It can be moved left, right, up, or down by various commands.

custom software: Software that has been written for a specific tailor-made purpose. It contrasts with packaged software that is written for general purposes.

data: The information, such as numbers or letters, put into the computer system.

data base: The large collection of related data that usually is in several files. It is generally accessible by the computer and thus is said to be on-line.

data base management system (DBMS): A software system/program that provides storage, access, updating, and maintenance of data in a data base.

data processing: Performing a programmed sequence of operations on a set of data to achieve a given result; also known as information processing. Commonly a term used to refer to the manipulation of large amounts of data by a computer.

DBMS: Data base management system.

debug: To remove the errors in a computer program.

deck: The collection of punched cards used in batch processing.

density: The distance between magnetic information on tapes or floppy disks. Higher density allows increased information storage capability.

device: A piece of hardware that performs some specific function: Input devices (e.g., keyboard) are used to get data into the CPU; output devices (e.g., printer or display monitor) are used to take data out of a computer in some usable form; input/output devices (e.g., terminal or disk drive) are able to perform both input and output of data.

diagnostic: A program for detecting and isolating a problem or mistake in the computer system; a feature that allows systems or equipment to self-test for flaws.

digital computer: A device that manipulates digital data and performs arithmetic and logic operations on these data. Digital data are represented in discrete, discontinuous forms as opposed to analog data, which are represented in continuous forms.

digital-to-analog converter: Device that transforms a computer's digital electrical pulses into a continuous analog signal to relay information to or power some nondigital device outside of the computer.

direct connect modem: A modem hard wired directly to the data transmission line, in contrast to an acoustic modem.

directory (or catalog): A list of the files stored on a peripheral storage device, like a disk. It can usually be obtained through the operating system program.

direct memory access (DMA): High-speed direct data transfer between an input/output channel and memory. This method provides transfers between a peripheral and the main memory.

disk: A revolving plate on which data and programs are stored.

disk drive: A part of the computer system that reads and writes material on the disk. It can be part of the main hardware or a peripheral attached to the system.

diskette: A small removable disk, often called a floppy disk.

disk operating system (DOS): A collection of programs that facilitates use of a disk drive and input/output of information between the computer and disk(s).

distributed processing network: The connections between a central computer and remote computers where data are transmitted to the central computer for complex processing and then sent back to the remote computer for review and further processing. This network is similar to timesharing in that it shares the cost and time of the expensive central computer.

DMA: Direct memory access.

documentation: (1) The instruction manual for a program or piece of hardware. (2) The process of describing a computer program so others using the program can see how it works.

DOS: Disk operating system.

dot matrix printer: A printer that uses a small array of dots to represent a coarse image of the characters printed. Most dot matrix printers that print upper-case characters use only a 5-by-7 matrix of dots to represent each character. Printers capable of upper- and lower-case printing usually use a 7-by-9 matrix of dots to represent a full set of alphabetic characters. High-resolution dot matrix devices like the inkjet or precision impact printers can assemble characters from matrices of 30-by-50 dots that may overlap, and are the ultimate in dot matrix technology.

download: To transfer a program or other data contained in a data base or bulletin board to a local computer. This process is performed while the local computer is connected to the distant computer via a telephone.

downtime: Any time a computer is not available or not working because of machine fault or failure. Downtime includes repair delay time, repair time, and machine-spoiled work time.

drill and practice: After a student logs on, the computer presents prescribed exercises and records the results. The instructor can retrieve detailed statistics on any student's progress.

driver: A program that provides an input format to an external device or another program. A printer driver receives input from the computer in the form of printed lines or graphic characters, and it outputs these instructions to a printer or plotter.

dumb terminal: An input/output device that does not use an internal CPU. Such devices require host computers for operation, whereas intelligent terminals have small internal central processing units to handle the terminal's functions and communications.

duplex: A method of operating a communications channel between two devices. Full duplex allows both units to send and receive simultaneously; half duplex allows only one unit to send information at a time.

EBCDIC: extended binary-coded decimal interchange code.

edit: To change or add data to an existing document or program.

editor: A program that manipulates text information and allows the user to make corrections, additions, deletions, and other changes.

electrostatic printing: A process in which an image is made on a suitable, special-purpose conductive paper by discharging a spark between the printhead electrode and the paper. The spark marks the surface layer of the paper by changing the appearance from a reflective silvery color to the dark color of the underlying layers of the paper.

end user: The individual who is the user of a product or service and does not provide or sell the product or service to anyone else.

ergonomics: The science of human engineering that combines the study of human mechanics and physical limitations with industrial psychology.

escape (ESC) key: A nonprinting key that causes the terminal and/or host to interpret subsequent characters differently.

execute: To run a program, using the instructions given.

extended binary-coded decimal interchange code (EBCDIC): Standard eight-bit code used to represent letters, numbers, and symbols.

external memory (external storage): Storage outside the computer in which information is retained even after the computer is turned off.

field: A group of consecutive characters that can represent a number, a word, a name, and so on.

file: A collection of data groups. Usually the groups in a file bear some relationship to one another.

firmware: The contents of a read-only memory (ROM) that are built into the chips or components of the computer in the factory and cannot be altered.

floating point BASIC: A form of BASIC language that allows the use of decimal numbers. Following calculations, the decimal point floats to a new position, as required, giving the term its name.

floppy disk: A small, flexible storage device made of magnetic material. It looks like a soft phonograph record and is usually 5¼ or 8 inches in diameter.

flowchart: A chart to show the sequence and branching of a particular procedure. This device is used frequently in the design of computer programs.

font: The set of images associated with a given character set like ASCII, EBCDIC, or the special-purpose sets used in computerized typesetting machines for magazines. A typical font for computer output from an impact printer might duplicate the font of a standard typewriter. For a low-resolution dot matrix printer, the font might be a program in the printer's read-only memory that translates each ASCII code into a visual representation as a matrix of dots.

forgive, forgiving: A characteristic of software that has been structured to allow a user to make minor errors in such a way that the system finds them and asks the user for the correct data or response.

FORTRAN (formula translation): A high-level computer language widely used to solve scientific and engineering problems.

friendly, user-friendly: Software characteristic implying that the software attempts to make data input/output easy and readily comprehensible by a user.

function keys: The keys on a keyboard used to perform operations defined by the user or program.

garbage: Meaningless information.

graphics: The technique of creating visual images by using a computer. Black-and-white or color television units are used for display with personal computers. The graphics can be used to show normal letters, numbers, and special symbols of a character set. Some personal computers have the ability to draw pictures.

handshaking: A method by which two different computer systems (or a computer and a peripheral device) can coordinate communication through some form of interconnection. A key part of this process is the ability to send messages about the status of the communications link, as well as messages that are part of the intended information.

hardcopy: A printout on paper of information from the computer.

hardware: All the physical parts of the computer system, including the computer itself, the input/output equipment and peripherals, and the physical disk or tape equipment.

hexadecimal: A number system that uses base 16 for its representation of integers. This base provides a more convenient, external, humanly readable representation of internal data. This base takes two digits from the set of numeric characters 0, 1, 2, 3, 4, 5, 6, 7, 8, 9, and the six letters, A, B, C, D, E, F, to represent the same number in binary form.

high-level language: A language, such as FORTRAN, BASIC, COBOL, and APL, that uses English language commands to keep the user from having to use machine code to communicate with the central processing unit. Typically, one high-level language statement will be equivalent to several machine-level instructions.

host computer: A computer that is the primary or controlling unit in a multiple-computer operation.

impact printing: A method of making a printed image by striking the paper in some way, usually involving a ribbon as in a standard typewriter. This method can use dot matrix character formation and sometimes uses predefined fonts as in the typewriter or on bands or chains of characters contained in some high-speed printers. This process is capable of producing multiple copies at the same time by using carbon paper or similar medium.

initialize: To set up the starting conditions necessary to run a program. To prepare a diskette or disk so that the computer can store data on it later.

inkjet printer: A device that prints characters by electrostatically aiming a jet of ink onto the paper.

input: The data put into the computer or the process of putting them in.

input device: A mechanism for entering information to the processor, usually a keyboard.

input/output device (I/O device): A peripheral device that has both input and output components, such as a video terminal. This device consists of channels (wires or telephone lines) within the computer system through which information flows. It also includes all the devices at the ends of wires or phones that originate or receive information. Some common I/O devices are card readers and punches, papertape readers and punches, typewriter devices, CRTs, magnetic tape, auxiliary disk systems, and line printers.

instruction: A group of bits that designates a specific computer operation.

integer BASIC: A form of BASIC in which only whole numbers can be processed (decimal numbers will not work).

integrated circuit: An electronic circuit or combination of circuits contained on semiconductor material, or a chip.

intelligent (disk, terminal, or other peripherals): Characteristic of a component that contains its own CPU so that it can execute instructions without the host's CPU.

interactive system: A computer system that allows two-way communication between the user and the computer.

interface: A piece of equipment used to connect two parts of a computer system that cannot interact directly with each other.

internal memory: Storage within the computer itself that is active as long as the computer is on.

interpreter: A computer language translator that translates and executes programs from a high-level language into a machine language, one line at a time.

I/O: input/output.

K: Abbreviation for kilo, meaning 1,000. In the term kilobyte, kilo stands for 1,024.

key in: To type or depress keys for data entry to a computer or storage medium.

keyboard: A group of buttons on a pad used to input information into a computer system.

keypunch: A typewriter-like keyboard device that punches holes (which represent data) in cards.

kilobyte: A measure of computer memory capacity. Each kilobyte of information is 1,024 bytes.

KSR (keyboard send receive): An indicator that the printer has a keyboard.

language: See **programming language.**

large-scale integration (LSI): The tens of thousands of microscopic electronic circuits and components crowded onto a square space measuring less than 1/8 inch on each side.

LISP (list processing): A widely used programming language in artificial-intelligence research.

listing: The actual lines of instruction making up a program.

load: To put data or programs into a computer.

LOGO: A program designed by Wallace Feurzeig at Bolt Beranek and Newman Inc., and named for the Greek word for "word" or "thought."

logoff or logout: The procedure for terminating a session at the computer.

logon or login: A sign-on procedure for beginning a session at the computer terminals of timesharing systems.

low-level language: A language easily understood by the computer. Low-level language programs are harder to write but faster to execute. They also take up less space than a program written in a high-level language. See **machine language.**

lpm (lines per minute): A unit of measure for the speed with which a printer produces copy, as in 300 lpm.

LSI: Large-scale integration.

machine language: The language that a specific machine was built to understand, written as a sequence of numbers. This language is immediately obeyed by the hardware, but is usually rather inconvenient to use.

machine readable: information stored on a peripheral storage device so that it can be recorded or played back to the computer.

magnetic tape: A recording device used to store programs and data. It resembles audio tape used in tape recorders.

main memory: A random-access form of memory that is the primary resource for storage of data and programs in a computer. Main memory is a temporary stage space in contemporary personal computers. When the power is shut off, the information is lost.

mainframe: Large computer systems. Mainframe systems with an expansive internal memory and fast processing time range in cost into the millions of dollars.

management information system (MIS): A system that, while performing routine processing functions, is designed to produce information for management to assist in decision-making. It often operates on large data bases.

mark sense card reader: An input device that can read cards that have information marked by graphite pencil.

mass storage: A technique that keeps track of large amounts of permanently available data in a machine-readable form. It is slower to access than main memory but yields larger potential amounts of data and has permanent qualities. Mass storage is provided by cassette tapes or floppy disks in most small personal computers.

MBASIC (MicroSoft BASIC): A version of the BASIC language used primarily on personal computers.

megabyte (mb): One million bytes of memory.

memory: That part of the computer that stores information. Also, the external material, such as floppy disks, hard disks, or cassette tapes that store information.

menu: The list of files and programs on a disk or tape.

microcomputer: A small computer in which the CPU is an integrated circuit deposited on a silicon chip.

microprocessor: A silicon chip that is the central, controlling part of the computer.

MicroSoft BASIC: See **MBASIC.**

minicomputer: A computer that is usually larger, more powerful, and more expensive than a microcomputer but with smaller memory and fewer functions than a mainframe computer.

mips: million instructions per second.

MIS: management information system.

mnemonic: Pertaining to the assistance of human memory: a mnemonic term, usually an abbreviation, that is easy to remember (mpy for multiply and acc for accumulation).

mode: A particular method of operation for a hardware or software device; for example, a particular unit might be capable of operating in a binary or a character mode.

modem (modulator/demodulator): A device used to attach a computer or one of its devices to a communication line, often a telephone.

modulator: An electronic black box used to translate the television output signals of the computer into a standard radio-frequency television signal that can then be fed into the antenna terminals of a television tuned to the appropriate channel. Usually on RF (radio frequency modulator).

monitor: A device to control or verify the operations of a system. It may also be referred to as a supervisor or an executive program and can be hardware or software. See **cathode ray tube.**

motherboard: A printed circuit board that has slots for various other circuit boards to be plugged into.

mouse: An electromechanical device connected to a computer by a cable. Movement of the device on a table top is mimicked by movement of the cursor on the computer screen.

MS-DOS: An operating system for personal computers; often contrasted with the CP/M operating system.

mtbf: Mean time between failures.

mttr: Mean time to repair.

networking: The sharing of resources or the communication between computers. See **communications network** and **distributed processing network.**

node: A junction or component in a network or an element in a tree.

noise: Any disturbance affecting the characteristics of a signal, such as random variations in voltage, current, or frequency. The term can also refer to data errors generated by disturbances in a circuit.

numeric pad: A keyboard for numeric input into a computer.

object code (object program): The machine-language form of a program, also called the object code of the program, that can be directly loaded into memory and executed since it has already been translated from its humanly readable form to the internally executable form.

octal: A numerical representation system where digits 0 through 7 are used to encode each of the eight possible sequences of 3 bits from 000 to 111.

on-line: A term referring to the location and connection of devices so that they are immediately accessible to the CPU of a computer. It also commonly refers to information that is directly obtained through a computer as opposed to that received through a book, television, and so on.

operating system: A special group of programs that controls the overall operation of a computer system. It mediates between the hardware and the particular software program.

operator: A symbol in a programming language that represents an operation to be performed on one or more operands (for example, + for add, × for multiply. Also the person who runs the computer.

optical scanner: An I/O device that reads clearly typed or printed information.

output: The information generated by a computer.

output device: A mechanism for showing information that has been "processed"; usually a printer or video display monitor.

packaged software: Software designed to be marketed for general use.

parallel: To handle simultaneously in separate facilities; to operate on two or more parts of a word or item simultaneously.

parallel interface: A method of connecting a peripheral device with a computer so that whole bytes (or groups of bytes) of data are transferred at one time. Multiple wires are therefore typically found in parallel interfaces. The parallel interface in a printer might include seven or eight data wires from three to five control wires. At the price of a more expensive connector, a much higher data transmission rate results. See also *serial.*

parity: A check bit or bits used to validate data.

Pascal: A compiled computer language, invented by computer scientist Niklaus Wirth (circa 1970), and initially intended as an aid to teaching computer languages. Its use is now widespread in computers of every size.

password: A safety device essential to protecting the privacy of a terminal user's programs. It is usually an alphanumeric word required to gain access to a personal storage area. This prevents interference, either accidental or deliberate, by unauthorized terminal users.

PC: Personal computer.

PEEK: An instruction in BASIC that enables the programmer to look at (peek at) any location in programmable memory. It is often used to scan the memory location that holds the information displayed on the video monitor to determine what is being displayed.

peripheral: A device, such as a CRT, a disk drive, or a printer, used for entering or storing data into or retrieving it from the computer system.

personal computer (PC): A small, low cost, portable computer with software oriented toward individual and small-business applications.

PFS: Preface to software packages made by one software company.

PILOT: A high-level language designed to make it easier for instructors to design software.

pinfeed: A standard feature of many computer printers that uses paper with holes along both edges to keep multiple page printouts in correct alignment.

pixel (picture elements): The smallest available unit of output in a graphics display device that can be controlled by the computer. In a dot matrix printer, the pixel is one dot within the matrix; on a television display device, the pixel is one dot on the screen of the television. Pixels can be black and white or colored, depending on the type of screen used.

PL/1: A high-level language developed to combine features required for commercial languages and scientific languages.

PLATO (programmed logic for automated teaching operations): This computer-based educational system originated at the University of Illinois in 1959. The system involves a very large computer with 4,000 student terminals that can be located anywhere in the world, with each terminal consisting of a key set and a plasma-display device. The plasma-display device is made up of two sheets of glass, forming words, pictures, or graphics. The glass is treated so that people can touch the screen to make responses with their fingers or a special pen.

POKE: An instruction in BASIC used to place a value (poke) into any location in programmable memory and often used in conjunction with PEEK.

port: The section of a computer through which peripheral devices can communicate.

printer: An output device to print information from a computer.

printout: The printed pages put out by a printer.

processor: An integrated circuit chip, or group of chips, that contains the electronic circuitry necessary for digital arithmetic and logical operations in a computer.

program: A set of coded instructions directing a computer to perform a particular function. Also known as software. Programs are written by programmers, and when a program is loaded into a computer, we say that the computer is programmed.

programmer: A person who converts a problem into a set of instructions for the computer in order to get a solution. See **program.**

programming: The design, writing, input, and testing of a computer program.

programming language: A special language of words and rules that is used to write programs so the computer can understand them.

PROM (programmable read-only memory): Memory that is not loaded with instructions during its fabrication but requires a physical operation to program. Some PROMs (EPROMS) can be erased and reprogrammed through special processes.

prompt: A message provided by an operating system calling for an action by the operator; also the symbol, ">," indicating such a message.

protocol: A computer communications procedure that includes the hardware of the interface and the software conventions of standard patterns that will be imposed on the data being communicated.

queue: A waiting line within the computer for use of a certain component. A queue develops most frequently in a time-sharing or resource-sharing system where several users need to use the same device.

RAM: Random access memory.

random access memory (RAM): The portion of the computer's memory in which data, instructions, and other information are stored temporarily. Also called read/write memory.

record: A collection of data groups. Usually the groups in a record bear some relationships to one another.

reliability: The measure of frequency of failure of the computer and other hardware.

remote access: Terminals physically at a distance from the central computer system (e.g., across town or across campus) at remote stations.

resolution: The quality of the image on a monitor as determined by the number of pixels on the screen, described by rows and columns. The larger the number of pixels is, the higher the resolution.

response (turnaround) time: The time interval between the request for a job to be done and receipt by the user of the results.

reverse video: A feature on a display unit that produces the opposite combination of characters and background from that which is usually employed; i.e., black characters on a white screen, if having white characters on a black screen is normal.

RO (receive-only): Refers to a printer that cannot be used to send information.

ROM (read-only memory): The portion of the computer's memory that contains information and instructions that are stored permanently. This memory cannot be altered or added to.

RS-232 (RS-232 interface): A data communications industry standard for the serial transmission of data to a peripheral device, like a printer, a video monitor, a plotter, and so on.

run: The continuous performance of the list of instructions in a given program or procedure. It is also the command to execute a program.

save: To store a program on a peripheral storage device for later use. It is also the command to do so.

scrolling: The movement of lines on a video display vertically in such a way that the top line disappears and a new bottom line comes into view at the bottom of the screen or vice versa.

semiconductor: A material such as silicon with a conductivity between that of a metal and an insulator. It is used in the manufacture of solid-state devices such as diodes, transistors, and computer chips.

serial: Handling of one after another in a single facility, such as transferring or storing in a digit-by-digit time sequence, or processing a sequence of instructions one at a time (sequentially). See also **parallel.**

sequential access: A storage method (such as on a magnetic tape) by which data can only be reached or retrieved by passing through all intermediate locations between the current one and the desired one.

simulation games: Games and representations of real-life situations. These are used when real-life equipment is too expensive or complex (e.g., cyclotron, nuclear reactor); measurement is impossible or disturbs the system (velocity of a falling body); experimental technique required is too complex (e.g., political promotion, science lab techniques); time scale is too long range (genetic studies, population dynamics, economic or social predictions); real-life situation is too dangerous (radiation from atomic reaction, explosive or toxic substances); or a problem requires extensive data collection and/or bookkeeping.

software: A general term for computer programs, procedural rules, and sometimes the documentation involved in the operation of a computer.

source program (source code): In a program written by a human being, the humanly readable form seen on the terminal that gets edited, changed, and updated in the process of creating a program. The translator program operates on the source language to produce the object code of the machine language.

spreadsheet: A class of computer programs designed to facilitate data manipulation in cellular arrays comprising rows and columns. Such a program looks much like an accountant's ledger sheet with the user inputting how a datum in one cell is related to those in other cells.

statement: A part of a computer program, an instruction that is not executed immediately but only after a command to the processor.

storage (memory): That component of the computer that retains information. Some typical forms of data storage are magnetic disks (flat spinning disks with magnetizable surfaces); magnetic drums, which hold more than 11 million bytes and take about 2.5 milliseconds to retrieve; and punched cards, a common and cheap storage medium, holding eighty letters or numbers.

string: A group of characters stored by their numeric codes, that are used in high-level languages such as BASIC.

subroutine: A program segment permitting a frequently required task to be called from any point of the main program. Execution is transferred to a subroutine when a subroutine call occurs. Subroutines save memory space at the expense of execution speed.

synchronous communication: A form of serial communication requiring common clock pulses between transmitter and receiver. The term synchronous indicates that events will occur with a regular or predictable time relationship.

system: The computer and all its related components, including hardware and software, that work together.

tape drive: A peripheral device for storing programs and other information on magnetic tape.

tapes: An inexpensive mass storage medium convenient for large files or archival storage but requiring sequential access.

telecommunications: The art and practice of sending computer (or verbal) messages through the telephone network or via radio. In personal computing, it refers to the use of serial communications techniques and modems that allow messages to be sent by telephone to other personal computers or to centralized information services.

teletext: Information in text form transmitted to people's homes through their television screens.

template: A plastic guide covering a keyboard with alternate key meanings for specific programs. The term also is used to designate the mathematical relationships between cells in a spreadsheet program as well as in stored text.

terminal: A peripheral device through which information is entered into or extracted from the computer, usually with a keyboard and an output device such as a CRT or printer.

terminal emulation: A communication method in which a terminal or a suitably programmed computer acts as a terminal of a particular design so that it can be used on various systems.

thermal printing: A method of scanning special heat-sensitive paper by moving a printhead that contains a dot matrix of electronically controllable heated areas. The heated zones are turned on if a dot image is to be recorded as part of the dot matrix representation of a character during the paper scan.

timesharing: A method by which more than one person can use a computer at the same time at separate terminals.

touch panel: A device sensitive to touch affixed to the front of the terminal display screen. It is used to input information at a particular screen location.

translator: A program capable of translating statements written in one programming language to the format of another.

turnaround time: See **response time.**

tractor feed: An attachment used to move paper through a printer. The roller that moves the paper has sprockets on each end that fit into the fanfold paper's matching pattern of holes.

turnkey system: A computer system that has all hardware and software installed. All a user supposedly has to do is turn it on.

turtle: The cursor on a screen using LOGO.

tutorial: CAI program that provides actual instruction. The computer "tells and asks" the student facts and questions and the teacher takes on the role of consultant or resource person.

UNIX: A small computer operating system developed by Bell Laboratories that features multiprogramming, a hierarchical file structure, and numerous useful utilities.

upload: The term used to describe data transfer on a connection between a distant computer with a data base or bulletin board and a local computer. It is the transfer of a program listing or other information from the local to the distant computer.

user-friendly: See **friendly.**

upload: See **distributed processing networks.**

utility programs: Support programs such as assemblers, editors, and sort, copy, and debug programs.

variable: A memory location in a computer language in which a character or a number may be stored. It usually has a symbolic name created by the person writing the program.

video terminal (video display terminal): A terminal that uses a video display unit like a monitor or CRT as its output device; a CRT plus a keyboard. See **cathode ray tube.**

videotext: An interactive data base search system utilizing telephones, computers, and television sets to make large computer data bases available for public access, usually by display-menu selection.

virtual memory: The amount of memory logically available to the program. It may be more than the physical memory size.

VisiCalc: A spreadsheet program marketed by Personal Software.

volatile: Characteristic of information that disappears from the memory of the computer when the power is turned off.

window: A section of a CRT screen that displays specific types of information; a split screen. In an integrated software package, a section of the display screen showing data from another part of the program. An example would be spreadsheet values being displayed in a section of the word processing display screen.

word: Ordered set of characters that occupies one storage location and is treated by the computer circuits as a unit and transferred as such.

word processing: typing, editing, storing, and printing text with a computer.

WordStar: A word processing package developed by MicroPro.

Appendix B: Agricultural Software Publishers

This list is not intended to be an all-inclusive review of agricultural software manufacturers. It was compiled from the files of the authors for the reader's convenience. No endorsement or recommendation of any company is intended or implied. If certain commercial suppliers or products are not listed here, it implies nothing other than the authors' lack of data regarding those suppliers or products.

Several commercial sources put out comprehensive lists of software suppliers. Some of these are:

Ag-Pros Microsystems (directory)
Box 64539
Lubbock, TX 79464

Agricultural Computing (newsletter with directory)
Doane Ag. Service, Inc.
8900 Manchester Rd.
St. Louis, MO 63144

Agricultural Software Directory (directory)
45 Dunfield Ave.
#1119
Toronto, ON M4S 2H4

Agriware Directory (directory)
3426 E. 242 St.
Elko, MN 55020

Farm Computer News (newsletter with directory)
c/o *Successful Farming Magazine*
1716 Locust
Des Moines, IA 50309

Commercial Software Companies

This list includes the company name and major software products it has advertised (where known).

Company	*Software*
ABC Systems, Inc. Box 475 Western, NE 68464	Accounting systems
Ag Data Systems 891 Hazel St. Gridley, CA 95948	General agricultural software
Ag-Plus Software 906 S. Main Ida Grove, IA 51445	Accounting systems
Ag-Ware P.O. Box 1503 Kearney, NE 68847	Accounting systems
Agricultural Computer Applications, Inc. P.O. Box 8, 1320 Notre Dame Dr. Davis, CA 95617	Least-cost rations systems
Agricultural Consultants, Inc. 124 E. Washington St. P.O. Box 711 Lebanon, IN 46052	Accounting systems
Agricultural Management Systems, Inc. 2770 Vista Avenue Boise, ID 83705	Shubox accounting system
Agricultural Software Consultants, Inc. 1706 Santa Fe Kingsville, TX 78363	Least-cost rations systems Feed pricing systems
Agricultural Software Systems, Inc. 1216 Dawson Rd., Suite 107 Albany, GA 31707	Bestcrop

Agri-Management Services P.O. Box 6425 Athens, GA 30604	Accounting systems General agricultural software
Agri-Data Resources, Inc. 205 W. Highland Ave. Milwaukee, WI 53203	On-line systems General agricultural software
Agrimanagement, Inc. P.O. Box 583 Yakima, WA 98907	General agricultural software
Ag-tronics Associates P.O. Box 221 Hyde Park, UT 84318	Irrigation management systems General agricultural software Soils conservation management systems Irrigation scheduling systems
Altamont Computers, Inc. 807 Central Ave. Tracy, CA 95376	Agstar (farm accounting)
A. O. Smith—Harvestor, Inc. Agristor Software Division P.O. Box 2000 Elm Grove, WI 53122	Dairy management system General agricultural software
Argos, Inc. 790 W. Shaw Ave., Suite 360 Fresno, CA 93704	Packer systems Irrigation systems General agricultural software
Arkansas Systems, Inc. 8901 Kanis Road Suite 201 Little Rock, AR 72205	Poultry systems
Briston Computers, Inc. Rt #3, Box 444C Albany, KY 42602	Agricultural and forestry systems
Brubaker & Assoc., Inc. 116 W. Main St. Delphi, IN 46923	Crop decision modes General agricultural software
Climate Assessment Technology, Inc. 11550 Fuqua St., Suite 355 Houston, TX 77034	Farm weather systems

Compco Computers, Inc. Division of Northouse Industries, Inc. 710 W. Fond du Lac Ave. Milwaukee, WI 53218	Dairy systems General agricultural systems
Computone Systems, Inc. One Dunwoody Park Atlanta, GA 30338	Elevator systems
Control Data Corporation Agricultural Products Division 1415 Energy Park Dr. St. Paul, MN 55111	Dairy systems Swine systems
Countryside Data, Inc. 718 N. Skyline Dr., Suite 201 Idaho Falls, ID 83402	Accounting systems General agricultural software
Crop Care Associates 1527 E. Shields Fresno, CA 93704	Crop management systems
Crop Data Management Systems 1521 Butte House Road Suite B Yuba City, CA 95991	Records, data base, and accounting systems
Cyberia, Inc. P.O. Box 784 Ames, IA 50010	Accounting systems General agricultural software
DHI Computing, Inc. Box 1427 Provo, Utah 84601	Dairy herd improvement and dairy systems
Dairy Herd Management Service, Inc. 715 South Bank Road Elma, WA 98541	Comprehensive dairy system
Data Craft Services P.O. Box 1048 East College Drive Marshall, MN 56258	Veterinary systems

Publisher	Software
Decision Data Services, Inc. 116 Kellogg Ave. Ames, IA 50010	General agricultural software
Delta Farm Systems, Inc. P.O. Box 2266 Grand Junction, CO 81502	Homestead accounting systems
Dickey-John Corporation Agricultural Management Division P.O. Box 10 Auburn, IL 62615	Swine systems Accounting systems
Efficiency Resources, Inc. 1101 H. Ave. Kalona, IA 52247	Swine systems
FBS Systems (Farm Business Systems) Box 201 Aledo, IL 61231	Accounting systems General agricultural systems Dairy systems Swine systems Crop records systems
Farm Management Systems of Mississippi, Inc. 223 3rd St. McComb, MS 39648	Accounting systems
Farmcomp 985 E. 6th So. Logan, UT 84321	Spreadsheet templates for a variety of computers
Farm Computer Systems Hillsboro, ND 58045	Farm ledger
FIS, Inc. (Farm Information Systems) P.O. Box 336 Waterproof, LA 71375	Accounting systems General agricultural software Grower mail order systems
Farmhand Computer Systems Box 40 Mitchellville, IA 50169	Swine management systems Accounting systems
Farm Management, Inc. 1208 South Cedar Road New Lenox, IL 60451	Financial systems

Farmplan Systems, Inc. 3130 Impala Drive Suite 200 San Jose, CA 95117	General agricultural software
Financial Systems, Inc. P.O. Box 2012 Kearney, NE 68847	Agricultural spreadsheet templates
Frontier Computer Corporation 2630 W. Durham Ferry Road Tracy, CA 95376	Farm records management system
Great West Seed, Inc. 610 W. Main Ave. Box 699 West Fargo, ND 58078	Accounting systems
Griffith Data Services 5410 Prospect Peoria, IL 61614	Accounting systems
Harris Technical Systems, Inc. 624 Peach St. Lincoln, NE 68508	Crop systems Swine systems Farm accounting systems General agricultural software
Harvest Computer Systems, Inc. 102 South Harrison St. Alexandria, IN 46001	Decision aids Accounting and budget systems Record keeping
Holm-Dietz Computer Systems, Inc. 555 W. Shaw, Suite C-14 Fresno, CA 93704	Agricultural management and crop costing systems Packer systems
IBM International Business Machines, Inc. Field Developed Programs Dept. P.O. Box 2150 Atlanta, GA 30055	Crop systems
Micro-Crop, Inc. 8245 Northwest 53rd St. Miami, FL 33166	Farm supplies management systems Fruit and vegetable systems

Modular Turnkey Systems, Inc. Route #7 P.O. Box 149 Hot Springs, AR 71901	Orchard/vineyard systems General agricultural software
Northwest Nutrition Management Consultants 916 South Columbia St. Milton Freewater, OR 97862	Cowpower systems
On Farm Computing 125 W. Market St. Suite 302 Indianapolis, IN 46204	Agricultural software
Pro-Farmer Professional Agricultural Software 219 Parkade Cedar Falls, IA 50613	On-line commodity information Pricing and charting systems
Profit Data, Inc. 808 Olena Ave. Willmar, MN 56201	Reaper accounting system
Radio Shack Division of Tandy Corp. 1800 One Tandy Center Fort Worth, TX 76102	Agricultural and other software
Range Management Software 1216 S. Ridgefield College Station, TX 77840	Grazing systems
Red Wing Business Systems, Inc. 610 Main St. Red Wing, MN 55066	Accounting systems Agricultural systems
Royalton Agricultural Service, Inc. Rt. 1, Box 43 Royalton, IL 62983	Agricultural software
Simplot Data Systems, Inc. 5383 Irving St. Boise, ID 83706	Accounting systems Potato grower systems

Smith, Dennis & Gaylord, Inc. 3211 Scott Blvd. Santa Clara, CA 95051	Broker systems
Small Business Computer Systems 4140 Greenwood Lincoln, NE 68504	Accounting systems Agri-Ledger
Specialized Data Systems, Inc. P.O. Box 8278 Madison, WI 53708	Accounting and spreadsheet systems
Stratsoft Serv. Corp. 2009 N. 14th St. Arlington, VA 22201	Stratacon Dairy system
Successful Farming Management Software c/o *Successful Farming Magazine* Locust at 17th Des Moines, IA 50336	General agricultural software
Summerville Enterprises, Inc. 104 Broad St. S. E. Aliceville, AL 35442	Accounting systems General agricultural software
Sysman, Inc. 6171 St. Rd. 26 West West Lafayette, IN 47906	Profit accounting system
TAG (Technical Analysis Group) P.O. Box 15951 New Orleans, LA 70175	Comprehensive commodity charting and analysis systems
Vertical Software, Inc. 502 E. War Memorial Peoria, IL 61614	Graintrac
Wisconsin Microware, Inc. 5201 Old Middleton Rd. Madison, WI 53705	General agricultural software

Appendix C: Microcomputer Periodicals

A+: Independent Guide to Apple Computing, Ziff-Davis Pub., 1 Park Avenue, New York, NY 10016

A.N.A.L.O.G. Computing, P.O. Box 615, Holmes, PA 19043

Access: Journal of Microcomputer Applications, P.O. Box 12847, Triangle Park, NC 27709

Access: Microcomputers in Libraries, Box 764, Oakridge, OR 97463

Agricomp, 1001 E. Walnut Street #210, Columbia, MO 65201

Analog Computing, 565 Main Street, Cherry Valley, MA 01611

Antic: The Atari Resource, 297 Missouri Street, San Francisco, CA 94107

Apple: The Personal Computer Magazine and Catalog, 20525 Mariani Avenue, Cupertino, CA 95014

Apple Assembly Line, P.O. Box 280300, Dallas, TX 75227

Apple Orchard, 910-A George Street, Santa Clara, CA 95050

Apple User, Database Publications, 68 Chester Road, Hazel Grove, Stockport SK7 5NY, Great Britain

Arithmetic Teacher, 1906 Association Drive, Reston, VA 22091

Auerbach Business Minicomputer Reports, 6560 North Park Drive, Pennsauken, NJ 08109

Auerbach General Purpose Minicomputer Reports, 6560 North Park Drive, Pennsauken, NJ 08109

Booklist, 50 East Huron Street, Chicago, IL 60611

Business Computer Systems, 221 Columbus Avenue, Boston, MA 02116

Business Computing, P.O. Box 815, Tulsa, OK 74101

Business Micro/Mini Reports, 140 Barclay Center, Cherry Hill, NJ 08034

Business Week, 1221 Avenue of the Americas, New York, NY 10020

Business Software, M&T Publishing Co., 2464 Embarcadero Blvd., Palo Alto, CA 94303

Byte: The Small Systems Journal, 70 Main Street, Peterborough, NH 03458

CAD/CAM Technology, 1 Sme Drive, Dearborn, MI 48128
Call A.P.P.L.E., 21246 68th Avenue S, Kent WA 98055
Classroom Computer News, P.O. Box 266, Cambridge, MA 02138
Collegiate Microcomputer, M&T Publishing Co., Rose-Hulman Institute of Technology, Terra Haute, IN 47803
Color Computing Magazine, New England Publications, Highland Mill, Camden, ME 04843
Color Computing News, P.O. Box 1192, Muskegon, MI 49443
Commodore Computing International, 193 Wardour Street, London W-1, England
Commodore: The Microcomputer Magazine, 487 Devon Park Drive, Wayne, PA 19087
Compute: The Journal for Progressive Computing, P.O. Box 5406, Greensboro, NC 27403
Computer Decisions, Hayden Publishing Co., 50 Essex Street, Rochelle Park, NJ 07662
Computer Equipment Review, 520 Riverside Avenue, Westport, CT 06880
Computer Graphics World, 1714 Stockton, San Francisco, CA 94133
Computer Law Journal. Law & Technology, P.O. Box 4658 T.A., Los Angeles, CA 90051
Computer Music Journal, MIT Press, 28 Carleton Street, Cambridge, MA 02142
Computer Talk for the Pharmacist, 1750 Walton Road, Blue Bell, PA 19422
Computer Talk for the Physician, 1750 Walton Road, Blue Bell, PA 19422
Computer Times: The Newspaper for the Minicomputer Industry & Small Systems User, 50 Essex Street, Rochelle Park, NJ 07662
Computer Update, Boston Computer Society, 3 Center Plaza, Boston, MA 02108
Computer Use in Social Service Network, P.O. Box 19129, Arlington, TX 76019
Computer User, 16704 Marquardt Avenue, Cerritos, CA 90701
Computers & Education, Pergamon Press, Maxwell House, Elmsford, NY 10523
Computers & Electronics, One Park Avenue, New York, NY 10003
Computers & Graphics, Pergamon Press, Maxwell House, Elmsford, NY 10523
Computers & Medicine, P.O. Box 36, Glencoe, IL 60022
Computers in Healthcare, 6430 S. Yosemite Street, Englewood, CO 80111

Computers in Mathematics and Science Teaching, P.O. Box 4455, Austin, TX 76765

Computers in Psychiatry/Psychology, 26 Trumbull Street, New Haven, CT 06511

Computerworld: Newsweekly for the Computer Community, 375 Cochituate Road, Box 880, Framingham, MA 01701

Compute's Gazette, Box 5406, Greensboro, NC 27403

Computing Teacher, 1787 Agate Street, Eugene, OR 97403

Computing Today, 145 Charing Cross Road, London WC2H OEE, England

Computronics Monthly Newsmagazines for TRS-80 Owners, 50 N. Pascack Road, Spring Valley, NY 10977

Core, P.O. Box 44549, Tacoma, WA 98444

Creative Computing: Magazine of Personal Computer Applications and Software, Ahl Computing, Inc., P.O. Box 789M, Morristown, NJ 07960

Data Cast: The User's Reference Publication for Major Microcomputer Software System, 345 Swett Road, Woodside, CA 94062

Data Communications, 1221 Avenue of the Americas, New York, NY 10020

Datapro Directory of Microcomputer Software, 1805 Underwood Blvd., Delran, NJ 08075

Datapro Directory of Small Computers, 1805 Underwood Blvd., Delran, NJ 08075

Datapro Reports on Minicomputers, 1805 Underwood Blvd., Delran, NJ 08075

Digit Magazine, 1 Park Avenue, New York, NY 10016

Digital Review, 1 Park Avenue, New York, NY 10016

Dr. Dobb's Journal for Users of Small Computer Systems, Box E, Menlo Park, CA 94025

Educational Computer, P.O. Box 535, Cupertino, CA 95015

Educational Technology, 140 Sylvan Avenue, Englewood Cliffs, NJ 07632

Electronic Learning, 902 Sylvan Avenue, Englewood Cliffs, NJ 07632

Enter, P.O. Box 2686, Boulder, CO 80322

Family Computing, 730 Broadway, New York, NY 10003

Forth Dimensions, P.O. Box 1105, San Carlos, CA 94070

Hardcore Computist, P.O. Box 44549, Tacoma, WA 98444

Home Entertainment Quarterly, 220 Westbury Avenue, Carle Place, NY 11514

Hot Coco, 80 Pine Street, Peterborough, NH 03458

ICP Directory. Mini-Small Business Systems: Cross Industry, 9000 Keystone Crossing, Box 40946, Indianapolis, IN 46240
ICP Mini-Small-Business Systems Software Directory, 9000 Keystone Crossing, Indianapolis, IN 46240
IEEE Micro, 345 E. 47 Street, New York, NY 10017
IBM Softalk, 11160 McCormick Street, North Hollywood, CA 91601
inCider: Green's Apple Magazine, 80 Pine Street, Peterborough, NH 03458
Infoage, Canada's Small System Magazine, 211 Consumers Road, Willowdale, Ont M2J4GH, Canada
Infoworld: The Newsweekly for Microcomputer Users, 530 Lytton Avenue, Ste. 303, Palo Alto, CA 94301
Instructional Innovator, 1126 16th Street NW, Washington, DC 20036
Interface: Computer Education Quarterly, 915 River Street, Santa Cruz, CA 95060
Interface Age: Business Oriented Microcomputer Magazine, 16704 Marquardt, Box 1234, Cerritos, CA 90701
International Journal of Mini and Microcomputers
(a/k/a *Mini and Microcomputers*), Box 2481, Anaheim, CA 92804
Journal of Computer Based Instruction, Western Washington University, Bellingham, WA 98225
Journal of Computers in Mathematics & Science Teaching, P.O. Box 4445, Austin, TX 78765
Journal of Microcomputer Applications, 24-28 Oval Road, London NW1 7DX, England
Journal of PASCAL & ADA, 898 S. State Street, Orem, UT 84057
Laboratory Microcomputer, 12 Clarence Road, Kew, Surrey TW9 3NL, England
Lawyer's Microcomputer, A Newsletter, P.O. Box 1046, Lexington, SC 29072
Lawyer's PC: A Newsletter for Lawyers, P.O. Box 1046, Lexington, SC 29072
Library Hi-Tech, P.O. Box 1508, Ann Arbor, MI 48106
Lifelines, The Software Magazine, 1651 Third Avenue, New York, NY 10028
Link-Up, 3938 Meadowbrook Road, Minneapolis, MN 55426
List, 3381 Ocean Drive, Vero Beach, FL 32963
MIMI. Represents International Society for Mini and Microcomputers, Symposium Proceed, Box 2481, Anaheim, CA 92804
Mathematics Teacher, 1906 Association Drive, Reston, VA 22091
Media & Methods, 1511 Walnut Street, Philadelphia, PA 19102

Micro: The 6502/6809 Journal, 34 Chelmsford Avenue Box 6502, Chelmsford, MA 01824

Micro, Advancing Computer Knowledge, P.O. Box 6502, Amherst, NH 03031

Microcomputer Applications, Box 2481, Anaheim, CA 92804

Microcomputer Newsletter: Application of Microcomputers to the Building Service Industry, Old Bracknell Lane West, Bracknell, Berks RG12 4AH, England

Microcomputer Owner, Box 41345 Craighall 2024, Transvaal, Johannesburg, South Africa

Microcomputer Printout, 7A Harpton Parade, Village Way, Yateley GU17 7TD, England

Microcomputer Systems D.A.T.A. Book, Box 26875, San Diego, CA 92126

Microcomputing, 80 Pine Street, Peterborough, NH 03458

Microdecision, 53-55 Frith Street, London W1A 2HG, England

Microprocessing & Euromicro Journal: International Journal of Microprocessing & Microprogram, Box 211, 1000 AE Amsterdam, Netherlands

Microprocessor I CS D.A.T.A. Book, Box 26875, San Diego, CA 92126

Microprocessors & Microsystems: Hardware, Software, Applications, Box 63, Westbury House, Bury Street, Guildford, Surrey GU2 5BH, England

Microprocessors at Work: Monthly Bulletin on Microprocessors & Microprocessor Application, Mayfield House, 256 Banbury Road, Oxford, OX2 7DH England

Microsystems: The CP/M and S-100 User's Journal, 39 East Hanover Street, Morris Plains, NJ 07960

Mini-Micro Computer Report: A Monthly Digest of Significant Trends in Mini-Comp & Micro-Proces, 7620 Little River Turnpike, Annandale, VA 22003

Mini-Micro Software, 322 St. John Street, London, EC1V 4QH England

Mini-Micro Systems, 221 Columbus Ave., Boston, MA 02116

Minicomputer News, 167 Corey Road, Brookline, MA 02146

Mininews, 1805 Underwood Blvd., Delran, NJ 08075

Mumps Users' Group Quarterly, 4321 Hartwick Road, #308, College Park, MD 20740

NYPC Newsletter, 1385 York Avenue, New York, NY 10021

Nation's Business, 1615 H Street NW, Washington, DC 20062

NEWS80S, P.O. Box 1329, Redondo Beach, CA 90278

Nibble: The Reference for Apple Computing, Box 325, Lincoln, MA 01773
North Star Notes, 14440 Catalina Street, San Leandro, CA 94577
OS/Tech, Box 517, Clearwater, FL 33517
PC, 1528 Irving Street, San Francisco, CA 94122
PC: The Independent Guide to IBM Personal Computers, 1528 Irving Street, San Francisco, CA 94122
PC: The Technical Journal, P.O. Box 598, Morris Plains, NJ 07950
PC User (St. Clair Shores), 25875 Jefferson at Madison, St. Clair Shores, MI 48081
PC World: The Personal Computer Magazine for Second-Generation IBM PCS and Compatibles, 555 De Hard Street, San Francisco, CA 94107
Peelings II, P.O. Box 188, Las Cruces, NM 88004
Personal Computer Age, 10057 Commerce Avenue, Tujunga, CA 91042
Personal Computer World, 62 Oxford Street, London W1, England
Personal Computers Today, 7315 Wisconsin Ave., Ste. 1200 N, Bethesda, MD 20814
Personal Computing, 50 Essex Street, Rochelle Park, NJ 07662
Personal Software, 50 Essex Street, Rochelle Park, NJ 07662
Pipeline, P.O. Box C, Oakdale, IA 52319
Pocket Computing, 70 Main Street, Peterborough, NH 03458
Popular Computing, 70 Main Street, Peterborough, NH 03458
Portable Companion, 26538 Danti Court 221, Hayward, CA 94545
Portable Computer, 500 Howard Street, San Francisco, CA 94105
Portable 100, Highland Mill, Camden, ME 04843
Power Play Computing, 1200 Wilson Drive, West Chester, PA 19380
Power/Play Magazine, 487 Devon Park Drive, Wayne, PA 19087
Pro-Files Magazine for Kaypro, P.O. Box N, Del Mar, CA 92014
Radio-Electronics, 200 Park Avenue South, New York, NY 10003
Rainbow, 5803 Timber Ridge Drive, Prospect, KY 40059
Reference (Amherst), Box 1200-A, Amherst, NH 03031
REMark, Hiltop Road, St. Joseph, MI 49085
SIGPC Newsletter, 1133 Avenue of Americas, New York, NY 10036
S-100 Microsystems, 39 E. Hanover Avenue, Morris Plains, NJ 07950
SATN: Journal for Visicalc Users, 27 Mica Lane, Wellesley, MA 02181
School Microwave Directory, Box 246, Dresden, ME 04342
School Microwave Reviews, Box 246, Dresden, ME 04342
Sextant, 716 E Street, SE, Washington, DC 20003

Sigsmall Newsletter, 1133 Avenue of the Americas, New York, NY 10036
Simulation, P.O. Box 2228, La Jolla, CA 92038
Simulation & Games, 275 S. Beverly Drive, Beverly Hills, CA 90212
Small Business Computer News, 140 Barclay Center, Cherry Hill, NJ 08034
Small Business Computers Magazine, Box 789-M, Morristown, NJ 07960
Small Computers in Libraries, 1515 E. First Street, Tucson, AZ 85721
Small Systems World, 950 Lee Street, Des Plaines, IL 60016
Softalk for the IBM Personal Computer, 11021 Magnolia Blvd., North Hollywood, CA 91601
Softline, 11021 Magnolia Blvd., North Hollywood, CA 91601
Softside, 6 South Street, Milford, NH 03055
Software News, Five Kane Industrial Drive, Hudson, MA 01749
Software Review, 520 Riverside Avenue, Westport, CT 06880
Spreadsheet, P.O. Box 254, Scarsdale, NY 10583
Sync, 39 E. Hanover Ave., Morris Plains, NJ 07950
Syntax Quarterly, Bolton Road, R.D. 2, Box 457, Harvard, MA 01451
Teaching and Computers, Box 2001, Englewood Cliffs, NY 07632
T.H.E. Journal, P.O. Box 992, Acton, MA 01720
The Scarlett Letter, 1301 N. 19th Street, Norfolk, NE 68701
Today Magazine, 500 Arlington Blvd., Columbus, OH 43220
Two/Sixteen + Twelve Magazine, 2216 1/2 W. James Street, Lancaster, PA 17603
UNIX Review, P.O. Box 563, Denville, NJ 07936
Videogaming Illustrated, 32 Oak Ridge Road, Bethel, CT 06801
Videoinfo: Video Markets & Technology News Bulletin, Box 3, Newman Lane, Alton, Hampshire GU34 2PG, England
68 Micro Journal, 5900 Casandra Smith, Hixon, TN 73743
80 Micro; Magazine TRS-80, 80 Pine Street, Peterborough, NH 03458
80-U.S. Journal, 3838 South Warner Street, Tacoma, WA 98409
99'er Home Computer Magazine, Box 5537, 1500 Valley River Dr., Suite 250, Eugene, OR 97405

Appendix D: Data Bases, Networks, and Service Groups

Agriculture-related Information Systems	*Product*
AGNET, University of Nebraska 105 Miller Hall Lincoln, NE 68583 402-472-1892	On-line information and problem-solving service for farmers, ranchers, agribusinesses, and homes.
AGRICOLA National Agricultural Library Beltsville, MD 20705 301-344-3755	On-line information service providing access to data on published literature pertaining to agriculture.
Agri-Data Resources, Inc. 205 West Highland Ave. Milwaukee, WI 53203 414-278-7676	On-line information on business, financial, marketing, weather, and price data. Also analyses and recommendations.
Alberta Agriculture 7000 - 113 Street Edmonton, Alberta Canada T6H-5T6	On-line information data base of agricultural commodities (prices/volumes). Access to AGDATA.
AMS Market News Network USDA, APD Washington, DC 20250 202-447-5115	Market News Telecommunications System reporting up-to-the-minute information on commodity prices, demand, and movement.

Chase Econometrics 150 Monument Rd. Bala Cynwyd, PA 19004 215-667-6000	Economic information and analyses on industrial economics, energy, fertilizer, minerals, international and U.S. economics, and agriculture.
Computerized Management Network Virginia Cooperative Extension Blacksburg, VA 24061 703-961-5184	Information system for solving problems, retrieving data, and evaluating programs.
CompuServe Inc. P.O. Box 20212 Columbus, OH 43220 614-457-8600	Private source offering access to agricultural news and the Commodity News Service data. Also provides shopping, banking.
Current Research Information System National Agricultural Library Beltsville, MD 20750 301-344-3850	Computerized information storage and retrieval covering U.S. publicly supported agriculture and forestry research.
Data Resources, Inc. 24 Hartwell Ave. Lexington, MA 02173 617-861-0877	Agriculture data bank and forest products service data bank; farm and industrial equipment (FIEI) data bank.
ESTEL University of Maryland Cooperative Extension Service College Park, MD 20742 301-454-4848	Provides current information on a number of topics by videotext equipment.
Instant Update 219 Parkade Cedar Falls, IA 50613 319-277-1278	Timesharing information delivery system designed for Professional Farmers of America. Offers a variety of services.

ITT Dialcom 1109 Spring Street Silver Spring, MD 20910 301-588-1572	Electronic mail system and information service. Provides national electronic mail support for Cooperative Extension Service.
National Electronic Marketing Assn. P.O. Box 722 Christianburg, VA 24073 703-382-1781	Computerized marketing systems for many agricultural products. Links buyers and sellers electronically.
NPIRS Purdue University W. Lafayette, IN 47907 317-494-8489	Nationally accessible on-line data base with information on all EPA-registered pesticides.
Source Telecomputing Corp. 1616 Anderson Rd. McLean, VA 22102 703-734-7500	World's first information utility, offering over 800 data bases and services. Provides access to more than 12,000 programs in many subject areas; a division of Reader's Digest.

References

Abbot, J. L. 1982. "Database Management with Ashton-Tate's dBASE II." *Byte* 7 (July):412–416.

Alpert, S. 1984. "Of Micros and Mainframes. *Computers and Electronics* 22 (May):12.

Barden, William, Jr. 1981. *Guidebook to Small Computers.* Indianapolis, Ind.: Howard W. Sams.

Barron, T., and G. Dieh. 1983. "Sorting Algorithms for Micro-Computers." *Byte* 9 (May):482–490.

Bartimo, J. 1984. "A 'Real' Computer in Your Lap: A New Generation of Portables Borrows Features from the Desktops." *Infoworld* 6 (May 7):91–94.

Bartimo, Jim. 1984. "The Art of Buying a Computer." *Infoworld* 6 (September 17):20–24.

Bartimo, Jim, and Scott Mace. 1985. "Winning Big by Thinking Small." *Infoworld* 7 (January):30–41.

Blundell, G. 1983. "Personal Computers in the Eighties." *Byte* 8 (January):166–182.

Borrell, Jerry. 1984. "Personal Computers Collide with High-Performance Terminals." *Mini-Micro Systems* 17 (November):133–138.

Bronson, Richard. 1984. "Computer Simulation: What It Is and How Its Done." *Byte* 9 (March):95–102.

Byers, T. J. 1984. "New Low-cost Modems." *Computers and Electronics* 22 (May):48–52.

Campbell, J. 1984. *The RS-232 Solution.* Berkeley, Calif.: Sybex.

Caruso, Denise. 1984. "Software Probes the Mind." *Infoworld* 6 (September 24):34–39.

———. 1984. "Modem Uses FM Radio Waves." *Infoworld* 6 (December 31):18–19.

Chang, D. 1983. "An Introduction to Integrated Software." *Byte* 8 (December):103–108.

Chin, K. 1984. "Firms Grab for dBASE II Gold." *Infoworld* 6 (April 9):72–73.

Ciarcia, S. 1983. "Keep Power-Line Pollution Out of Your Computer." *Byte* 8 (December):36–44.

Clapp, D. 1984. "Trial by Hard Disk." *Infoworld* 6 (June 18):32.

Cook, Rick. 1984. "A Machine for All Software." *Popular Computing* 3 (November):64–72.

"Delineating IBM-PC Compatibility." 1984. *Byte* 9 (April):273.

Falvey, Jack. 1983. "Real Managers Don't Use Computer Terminals." *Wall Street Journal*, February 7, p. 22.

Ford, Jerry D. 1984. "Four Educational Concerns in Using Microcomputers." *Teacher Education* 19 (3):16–20.

Freiberger, Paul. 1984. "The Videodisc Connection." *Popular Computing* 3 (September):64–71.

Friedman, Herb. 1984. "Don't Get Stuck." *Radio Electronics* 55 (April):61–63.

Gabel, D. 1983. "Word Processing: Finding the Right Software." *Personal Computing* 7 (April):110–130.

Gens, F., and C. Christiansen. 1983. "Could 1,000,000 IBM Users Be Wrong?" *Byte* 8 (November):135–141.

"Giving Novices Courage to Face Their Computers." 1982. *Business Week*, November 1, p. 85.

Glidewell, Richard. 1984. "New Printers Strive for Speed and Letter Quality." *Business Computer Systems* 3 (December):107–112.

Gugliotti, J., and E. Weitz. 1984. "Getting Mainframe Data To Micros." *Computers and Electronics* 22 (May):61–66.

Guide to Personal Computing. 1982. Bedford, Mass.: Digital Equipment Corp.

Hall, Mark. 1984. "Voice Input and Output." *Micro Communications* 1 (September):16–23.

Heintz, C. 1982. "Buyer's Guide to Word Processing Software." *Interface Age* 7 (December):40–58.

———. 1983. "Buyer's Guide to General Ledger Software." *Interface Age* 8 (July):70–84+.

Howitt, Doran. 1984. "Boom for Voice Mail Systems—Expansion Cards Turn Micros into Intelligent Answering Machines." *Infoworld* 6 (October 29):37–38.

———. 1984. "The Source Keeps Trying." *Infoworld* 6 (November 5):59–64.

———. 1984. "Photos to Digital Pictures Using a Personal Computer." *Infoworld* 6 (November 26):43–45.

Hungate, Lois A., and Ralph W. Sherman. 1979. *Food and Economics*. Westport, Conn.: AVI Publishing Co.

"IBM-PC Dominance Proclamation Will Prove Premature." 1984. *Infosystems* 31 (April):29.

Isaacs, A. 1984. "Compatibility: Software Compatibility and Standards." *Radio-Electronics* 55 (March):71–74.

Lachenbruch, P. 1983. "Statistical Programs For Microcomputers." *Byte* 8 (November):560–570.

Litzenberg, Kerry K. 1983. "Computer Use in the Agricultural Economics Classroom." *American Agricultural Economics Association* 64:970–987.

Mace, Scott. 1984. "Apple IIe Sales Surge as IIc is Readied." *Infoworld* 6 (April 9):54–55.

———. 1984. "Apple IIc Isn't Selling Out." *Infoworld* 6 (June 18):11–12.

———. 1984. "Users Groups Reach Out." *Infoworld* 6 (September 17):20–24.

———. 1984. "Computer Games Get Social." *Infoworld* 6 (October 15):30–31.

Mackenzie, Leonard N. 1983. "Teaching Old Dogs New Tricks: The Need for Adult Computer Literacy." *Vital Speeches of the Day* 50:58–61.

Marks, G. A. 1984. "Apple's Macintosh PC: Some Personal, Preliminary Observations—The Macintosh versus the IBM-PC." *ICPSR Micronews* (University of Michigan) 1, no. 2 (February 1984):1.

Marx, A. 1983. "Matchmaker! Matchmaker! Interfacing with Serial and Parallel Ports." *Computers and Electronics* 21 (November):74–79.

Mateosian, R. 1984. "1984: The Year of the 32-bit Microprocessor." *Byte* 9 (January):134–150.

McCredie, John W., ed. 1983. *Campus Computing Strategies.* Bedford, Mass.: Digital Equipment Corp.

McGeever, Christine. 1984. "On-Line Enthusiasm." *Infoworld* 6 (September 10):41–42.

McMullen, B. E., and J. F. McMullen. 1983. "The Super Spreadsheets: How Do They Compare?" *Popular Computing* 2 (June):112–123.

Meilach, D. Z. 1982. "Ten Steps to Take Before You Buy a Computer." *Interface Age* 7 (June):66–69.

Miller, F. W. 1984. "Smaller and Faster: IBM's Experimental Memory Chip Achieves Unprecedented Speed." *Infosystems* 31 (April):60–61.

Monk, J. Thomas, and Kennith M. Landia. 1983. "The Network that Beefed Up the Dairy." *Business Computer Systems* (December):35–36.

Naisbitt, John. 1982. *Megatrends: Ten New Directions Transforming Our Lives.* New York: Warner Books, Inc.

Osborne, A. 1979. *An Introduction to Microcomputers: The Beginners Book.* Vol. O. Berkeley, Calif.: Osborne.

Ouchi, G. I. 1984. "Lotus in the Lab." *PC World* 2 (February):222–229.

Pearlman, Dara. 1984. "The Joy of Telecomputing." *Popular Computing* 3 (July):107–110.

Popenoe, C. 1984. *Book Bytes: The User's Guide to 1200 Micro-computer Books.* New York: Pantheon Books.

Pournelle, J. 1983. "The Next Five Years in Microcomputers: Our User Unlocks His Crystal Ball and Becomes a Seer." *Byte* 8 (September):233–244.

Powell, David B. 1984. "Buyers Guide to Communications Software." *Popular Computing* 3 (July):121–126.

———. 1984. "Buyers Guide to Modems." *Popular Computing* 3 (July):111–117.

Pritchard, William H., Jr., and Donald Z. Spicer. 1983. "The Vassar Computer Literacy Program." *Educom* 18:2–3.

Quinn, James; Joseph Kirkman, and Cora Jo Schultz. 1983. "Beyond Computer Literacy." *Educational Leadership* 41(1):38–39, 67.

"REPORT CARD: ABSTAT Statistical Program from Anderson-Bell." *Infoworld,* Special Publication, 5 (December 1):50–52.

Segal, H., and J. Berst. 1982. *How To Select Your Small Computer Without Frustration.* Englewood Cliffs, N.J.: Prentice-Hall.

Shaw, Donald R. 1981. *Your Small Business Computer: Evaluating, Selecting, Financing, Installing and Operating the Hardware and Software That Fits.* Florence, Ky.: Van Nostrand Reinhold Co.

Smith, Dorthea M. 1985. "Searching with In-Search." *Micro Communications* 2 (January):27–29.

Spinrad, Norman. 1984. "Home Computer Technology in the 21st Century." *Popular Computing* 3 (September):76–83.

Stifter, F. 1983. "Coping with Power-Line Disturbances." *Computers and Electronics* 21 (October):35–43.

Strain, J. R., and S. Fieser. 1982. *Updated Inventory of Agricultural Computer Programs.* Gainesville, Fla.: University of Florida Cooperative Extension Service.

Summers, Tan. A. 1984. "The Computer Controlled Home." *Popular Computing* 3 (September):85–88.

U.S., Department of Agriculture (USDA). 1982. *The Computer: Management Power for Modern Agriculture.* Washington, D.C.: USDA Extension Committee on Organization and Policy.

______ . 1984. *Computers on the Farm.* USDA Farmer's Bulletin no. 2277. Washington, D.C.: U.S. Department of Agriculture.

U.S., Department of Agriculture Extension Service (USDA-ES). 1983. *How to Shop for a Home Computer.* USDA-ES Fact Sheet, Home Economics and Human Nutrition. Washington, D.C.

Watt, Dan. 1984. "Musical Microworlds." *Popular Computing* 3 (August):91–94.

Wortman, Leon A. 1985. "Wordstar 2000." *Infoworld* 7 (January):47–49.

Zaks, R. 1981. *DON'T: Or How to Care For Your Computer.* Berkeley, Calif.: Sybex.

Zientara, M. 1984. "PCJr Sales Below Expectations." *Infoworld* 6 (April 9):13–14.

Index